ALGEBRA EXAMPLES

BASIC FUNCTIONS 1

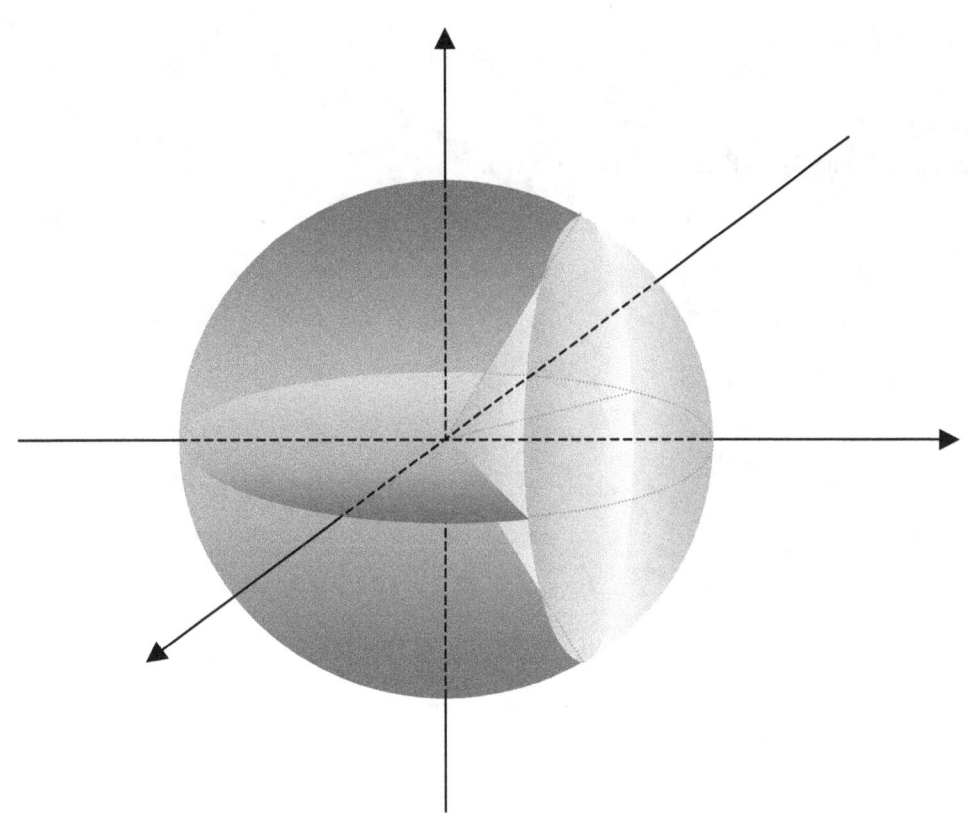

Seong R. Kim

Dear students:

Students need the best teacher, so you need examples, because examples are the best teacher. All the examples here are fully worked, and explain **how** the basic and essential tools in math are made, together with **what** they are, **how** they work, and **how** to work with them. Such tools include numbers, formulas, identities, equations, laws, etc.

Examples here begin with easy ones, of course. Covering every meter and yard properly, we can cover thousands of miles and kilometers. And it is particularly the case in math.

Of those examples therefore, some might even look too easy for you. It's not that easy though, to come up with those examples. Anyways, the bigger and the taller the tree, the deeper and the stronger the root.

Doing math, we work with ideas and run ideas, because every thing in math is an idea. A number is an idea, for instance, and the same is true for a line or circle, too. And putting ideas together, we build another, which becomes the base or an element of another, and each is connected. And that's the way your math grows. So you get to build a circuit, and sometimes, need to fill the gap or repair the circuit so that you get the sense of it.

So your calculation runs properly, and you get the problem solved.

The examples have been made and arranged so that they get tougher (or sometimes easier for some reason) as you proceed with them. In particular, similar examples with some variations are strategically repeated so that you can get the ideas or the tools tricky or complicated, and can get them mastered.

This book is however, nothing but a bunch of examples until you get it powered.   How then, to get it powered, and make it run and work for you?

Just read it, and then, do each example in writing. And it is important to note that you do it in **your** writing. Just watching someone doing it, you just only feel that you can do it. If you do it, you can do it, but if you don't, we can hardly. It's a cliché, of course, but is always true that knowing is one thing and doing is another.

I've been helping students grow, take care of, and run their own math. The area covers algebra and geometry for high school or college students, and is especially for equations (for unknowns or curves), functions, and their graphs, which are the basic elements in calculus, which's been the core of my interest from my early age in high school.

Of my students, some are quite poor in math, and thus, are afraid of or hate math, some require special education because of exceptional intelligence, some are smart enough, some are naïve and diligent, some are clever but lazy, and most behave in general. All the students are badly after though, one thing in common: a strong and secure math skill. It is of course, the prime objective of my work, and I'm always happy to and eager to help them achieve it. The problem was however, that many of them wanted it to be purchased. And the question is, can we buy it?

We can buy the means, of course. And a solid math skill is feasible, too. We know however, we can't buy love, and the same is true for the math skill, too. It's not what we can buy or sell, and not what we can give or take. It is however, what we can grow, and need to grow. Your math grows as much as you grow and take care of it. So does mine.

What math then, do students most often do or use in high schools or colleges?

It is algebra and geometry. What algebra though?

Elementary algebra, of course
Doing the algebra, we work with numbers (many in kinds), constants, variables, ratios, rates, expressions, equations, inequalities, functions, identities, formulas, laws, etc., together with signs and symbols. And if we want to do algebra properly, we want to know their natures and how they mingle with each other.

So studying math ideas or tools, you want to know **what** they are, **how** they work, and **how** to work with them or **what** to do with them.   What then, about the geometry?

Basically, the geometry has much to do with shapes, positions, and angles. The shapes begin with triangles and circles, and move on to rectangles, squares, parallelograms or rhombuses, trapezoids, tetragons, other polygons, polyhedrons, etc.

Doing the geometry, too, though, we need to do the algebra stated above. So it is analytic geometry, often called coordinate geometry, too. And doing it, we can specify positions using coordinates. So in the geometry, basically, we work with graphs. Putting a math idea in a graph, we can not only effectively think about it but actually see it, too, and therefore, can efficiently work with it. What idea then, is it?

The idea begins with a point, line, parabola, circle, ellipse, and hyperbola, called a conic section or basic curve, and then, moves on to other curves, planes, surfaces, volumes, and other objects in various dimensional spaces, together with vectors.

And using an angle, we can specify an amount of turn or change in direction.

So learning, using, or applying those ideas or math tools, we get to solve problems.

And this book can help. It can help learn them, and use them so that you can navigate to find solutions to problems. And in particular, it can help come up with answers to those **what**s and **how**s stated above. So it can help you grow and run your own math, and thus, can help achieve your solid math skill.

It is however, not a magic book giving you a math skill of high caliber overnight. And it can have many mistakes, too. There is no magic, and math is full of facts and ideas. And it is after all, not me and not your teacher but you who put together some of those facts and ideas, and understand it. Putting facts and ideas together, understanding it, and taking care of what you have learned, you grow your math. And this book can help.

This is a book of examples designed to help you grow your math, and assumes that you are a real beginner. This book requires though, time and effort, the amount of which need to be substantial, too, but will be worth it. That's because you want a substantial achievement, and will get it. And probably, you will get to see this book helping you get there much faster than expected. And then, you will get to see the way math runs.

In math, everything is an idea. So is a problem. And solving it, we put it many different ways. For instance, while expanding or reducing it, or modifying or converting it, we keep searching for the solution, approaching the solution, and eventually, can get there. So don't look for the solution outside the problem. The solution is inside the problem if the problem is properly made.

If it is not, no solution is the solution. And in fact, it is often the case a problem itself is the solution. We can put a problem in many different ways, and eventually, can end up with the solution.   How come then, is the solution no other than the problem?

For instance, the solution to $3232 \div 101$ is 32.   And we can put it this way:

$$3232 \div 101 = \frac{3232}{101} = \frac{32 \times 101}{101} = \frac{32}{1} = 32 \implies 3232 \div 101 = 32.$$

And we can get this, too: $32 \implies 3232 \div 101$.   How?

$$32 = \frac{32}{1} = \frac{32 \times 101}{101} = \frac{3232}{101} = 3232/101 = 3232 \div 101.$$ Too easy?

For another instance, the solution to $ax^2 + bx + c = 0$ is: $x = \frac{-b \pm \sqrt{b^2 - 4ac}}{2a}$, which is called the quadratic formula.   How come then, is the solution no other than the problem?

We can put it this way:

$$x = \frac{-b \pm \sqrt{b^2-4ac}}{2a} \implies 2ax = -b \pm \sqrt{b^2 - 4ac} \implies 2ax + b = \pm \sqrt{b^2 - 4ac}$$

$$\implies (2ax + b)^2 = b^2 - 4ac \implies 4a^2x^2 + 4abx + b^2 = b^2 - 4ac$$

$$\implies 4a^2x^2 + 4abx = -4ac \implies ax^2 + bx = -c \implies ax^2 + bx + c = 0.$$

And we can get this, too: $ax^2 + bx + c = 0 \implies x = \frac{-b \pm \sqrt{b^2-4ac}}{2a}$.   How?

$$ax^2 + bx + c = a(x^2 + \tfrac{b}{a} x) + c = a(x^2 + \tfrac{b}{a}x + \tfrac{b^2}{4a^2} - \tfrac{b^2}{4a^2}) + c = a(x^2 + \tfrac{b}{a}x + \tfrac{b^2}{4a^2}) - \tfrac{b^2}{4a} + c$$

$$= a(x + \tfrac{b}{2a})^2 - \tfrac{b^2-4ac}{4a} = 0 \implies a(x + \tfrac{b}{2a})^2 = \tfrac{b^2-4ac}{4a} \implies (x + \tfrac{b}{2a})^2 = \tfrac{b^2-4ac}{4a^2} \implies x + \tfrac{b}{2a} = \pm\sqrt{\tfrac{b^2-4ac}{4a^2}}$$

$$\implies x = -\tfrac{b}{2a} \pm \tfrac{\sqrt{b^2-4ac}}{2a} = \tfrac{-b \pm \sqrt{b^2-4ac}}{2a} \implies x = \tfrac{-b \pm \sqrt{b^2-4ac}}{2a}.$$

And we call the set of processes above, algebra.

So if a problem is well defined, that is, if it makes sense, we should be able to get it solved the way below:

A problem $\Rightarrow$ ... $\Rightarrow$ ... $\Rightarrow$ the solution, and thus: **the problem $\Rightarrow$ the solution**.

So solving a problem, we put it many different ways so that we can get to the solution.

And that's the way, math runs.

May your math run very well.

Seong R. Kim

B.S. Math. Michigan Tech. Univ.   M.S. Math. Rensselaer Polytechnic Institute

Notes:

This book is about an idea called functions. Why functions, though?

Using functions, we can see how things change.
More specifically, how values change as other values change.   How come?

Basically, functions are about values that change.
Expressing values, we use numbers, of course.
Not only that, of course. But we use letters, too, called variables or constants.
And using those letters, along with numbers, we make expressions that express values.
So values of expressions change as values of variables change.
And we want to know how their values change as values of their variables change.

And we call such an expression a function. So we want to know how the value of a
function changes as the value of the variable changes.
And thus, doing physics, economics, or anything that have to do with changes, we want
to know about functions. In short, how things change.

Increase or decrease
Going up or coming down
Speeding up or slowing down

So how fast or slow is it? And by how much or how many?

For instance:
How the distance changes as time changes.
How the pressure changes as temperature changes.
How the price changes as supply decreases or demand increases.

And after all, how outputs change as inputs change.
Functions get inputs as time, and in return, produce outputs as distance.

So if need to work with things change, you want to know functions, and how they work, together with how to work with them.

And functions are expressions mainly made of polynomials, made of variables, constants, and numbers, of course. So you want to know polynomials, how they work, and how to work with them, too. And as stated above, polynomials are expressions made of variables, constants, and numbers, of course. You want to know thus, variables and constants, how they work, and how to work with them, along with their arithmetic.

And this book covers all the basics and ideas stated above. What then, about numbers?

A function can use numbers of all kinds. Or we can say that we can use in a function number of all kinds. Mainly though, we use numbers said to be real. So at least, you want to know real numbers of all kinds, and we often put them into three groups: integers, rational numbers, and irrational numbers.

And we have numbers that have different looks. Among those, we often use powers, and have logarithms, called logs, too. And that's not it. We have radicals, called roots, too. And using those above, we can make expressions called functions. So you want to cover quite a few to learn, make, and use functions. So this book covers the basics and ideas using all such numbers when working with functions.

And we can actually see how functions behave, too. Each function can have its own look, which is called its curve, called a graph, too. And producing the graph or curve of a function, we can say we put the function in a graph or just we graph it. So putting the a function in a graph or graphing a function, you can see how the function behaves, and can find readily and fast the one you are busy looking for. And the one is of course, the solution to the problem you are busy doing. So this book shows many examples of graphs of many kinds.

Besides, though this book, you can improve your skill of algebra, too.
So the book does not just explain the things stated above. But it also helps follow steps to the solutions, and thus, helps you do calculations with expressions so that you can improve your calculation work when working with functions.

With strong skill of algebra, you can do a lot, and of course, can do problems very well, too. And through this book, you can grow much of your power in algebra.

And all the basics and many ideas stated above are covered in two separate books. And the two books are as follows:

ALGEBRA EXAMPLES BASIC FUNCTIONS 1

ALGEBRA EXAMPLES BASIC FUNCTIONS 2

And all the contents in the two books above are put together in one book as follows:

ALGEBRA EXAMPLES BASIC FUNCTIONS

So this book will help you learn what functions are about, how they work, and what to do with them, together with how to do it. And also, this will help improve your skill of algebra, too. You will soon be thus, able to change or alter, convert, or modify math expressions so that you can get to the solutions fast. And you will learn them through examples detailed so that your math can run not only properly but fast enough, too.

Contents

In BASIC FUNCTIONS 1

The Preview of the Contents

In BASIC FUNCTIONS 2

Note:

The drawings or graphs in this book are not exact, and are approximate or conceptual ones.

$\in$	"$a \in B$" means that a belongs to B. "p, q, and $r \in W$" means that p, q, and r belong to W.
$\Rightarrow$	"$A \Rightarrow B$." means that A implies B.
$\equiv$	$A \equiv B$ means that A and B are identical to each other.
$\neq$	$A \neq B$ means that A is not equal to B.
$\lvert A \rvert$	The magnitude of A. For instance, $\lvert -1 \rvert = \lvert 1 \rvert = 1$.
$\therefore$	Therefore
$\Leftrightarrow$	"$A \Leftrightarrow B$" means "If A then B." and "If B then A." We can read $A \Leftrightarrow B$ as "A if and only if B." In such a case, we can say that $A = B$.
Δx and Δy	Suppose that (x_1, y_1) and (x_2, y_2) are two points in the x-y plane. Then, we get either of the two below. $\Delta x = x_2 - x_1$, and $\Delta y = y_2 - y_1$. $\Delta x = x_1 - x_2$, and $\Delta y = y_1 - y_2$.

Distance Formula

Suppose that d is the distance between two points (x_1, y_1) and (x_2, y_2) in the x-y plane. Then, we get $d^2 = (\Delta x)^2 + (\Delta y)^2$.

0.0. **What is a function?**

Though it looks like an equation, it is not.   What then, is it?

- It's a *mathematical expression*. It's not just a math expression though. We can call it an expression that creates a *correlation between data sets*.   What data sets though?

The functions covered in this book are basic ones, which are thus covered in high school math or in the first year in college or university math.

So such data sets are simple sets of numbers, usually real numbers.
And in elementary level as high school math, we normally use two sets of numbers.

And usually, we use as a function a math expression as $2x$, $x^2 + x + 2$, 3^{x+2}, $\log x$, $\sin x^2$, etc. It can be of course, as simple as 1, 3, -2, or 0, since a number is an expression, too.

How then, does it create a correlation?

We can put a function this way, too:

- A function is an expression *acting on a number set to produce another number set*.

How does a function then, act on a number set?

A number from the set acted on is put into the variable in the function's expression, so a number gets produced. And the number produced belongs to the other set.

Suppose for instance, X is the set acted on, and is a set {1, 2, 3}, the function is: $3x + 1$, and Y is the set produced, and is a set {4, 7, 10}.

Then, if for instance, 1 from X is put into the expression $3x + 1$, we get: $3 \cdot 1 + 1 = 4$, so 4 gets produced. And thus, 4 belongs to Y.

And the same is true, too, for all the other numbers in both sets.

That is to say that a function connects two number sets.
One is the set acted on, and the other is the set of all the numbers produced.

A function is therefore, not just an expression, but creates a *system*, where two sets of numbers are connected by the function, which is an expression as $2x$ or $x + 1$.

   • In short, a function is an expression connecting two number sets.

And assuming F is a function, we can put it the way as follows:

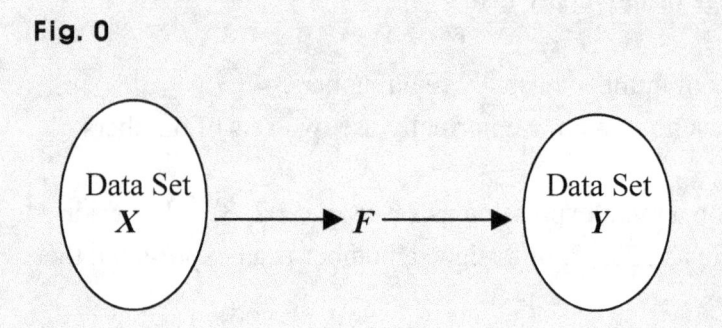

Fig. 0

Through the function F, the data set X gets connected to the data set Y.

Such two data sets are two sets of numbers. So a function creates a correlation between two number sets.   What correlation, though?

A function creates a correlation by setting up two relationships.

One is a *dependence* relationship between the two sets.

And the other is a *mechanical* relationship between the numbers in one set and those in the other set. What mechanical relationship?

Basically, the mechanical relationship goes the way below:

A pair of two numbers gets generated *at a time*. And the two numbers in each pair get connected. So number-pairs keep getting made as the function keeps running.

In each pair, one is from one set, and the other belongs to the other set. And the two numbers get connected by the function, because the function uses one to make the other.

• That is, the function connects two numbers by making a number-pair as (2, 7).

And a function can be said to generate a sequence of pairs of numbers.
That is, a function can be taken as a machine generating pairs of numbers in a sequence.

And in high school math and calculus, all the numbers in both sets have to get connected. Thus, the function keeps running until all the numbers get connected.   So?

So eventually, the function connects the two sets.
What then, about the dependence relationship?

Both sets can be the same or different. One can be bigger or smaller than the other. And it is often the case, both are identical. In nature though, the two sets are always different. How then, the two are different?

• One of the two is called a *domain*, and the other is called a *range*.
And the range is said to be dependent upon the domain.   How is it dependent, though?

We know a function is an expression acting on a number set, and producing another set of numbers. The *set acted on* is the *domain*, and the *set produced* is the *range*.

A number called an *input* from the domain is put into the variable in the function, which therefore, produces a number called an *output*, which belongs thus, to the range.

And the same is true, too, for all the other numbers in both sets.

So a number in the range can be said to get determined by a number in the domain. And the same is true, too, for all the other numbers in the domain and the range.

That is, if no number is chosen in the domain, no number can be determined in the range, and if a number is chosen in the domain, one number has to be determined in the range.

- So each number in the range is dependent upon at least one number in the domain.
- In sum, the range depends on the domain.   So no domain, then no range.

And for the same reason, we can say that the domain causes the range. And each number in the range can be said to *correspond to* at least one number in the domain.

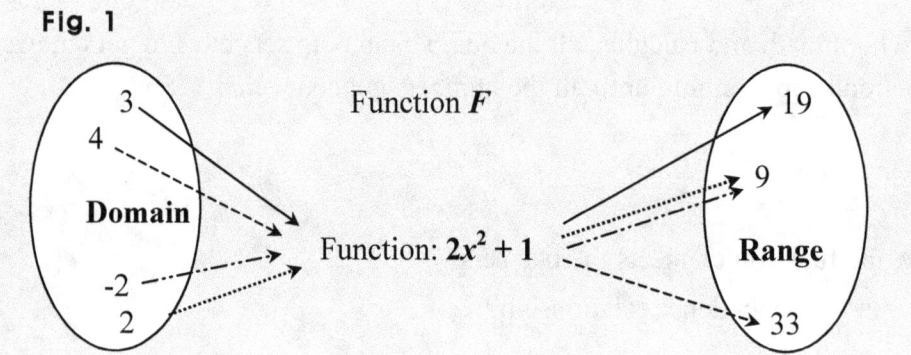

Fig. 1

So we can say that the numbers in the range correspond respectively to the numbers in the domain. And in the system that has the function *F* above, we can say that:

- 19 in the range corresponds to 3 in the domain.

- -2 gets chosen in the domain, enters the function, the function runs, then 9 gets determined in the range. And the same is true for 2 in the domain, too.

- Also, 4 in the domain is connected to 33 in the range, that is, the two get connects by means of the function: $2x^2 + 1$.

In sum, the domain and the range get connected via the function *F*, which is: $2x^2 + 1$.

And as others do in math, a function has its name.

And naming a function, we usually use two letters, together with brackets or parentheses.

One is called a function *designator* (another word for a name), and the other is the variable used in the expression.

So for instance, if a function is called f, and the expression is: $2x + 1$, we can name the function this way: $f(x)$, read as f of x.

Then, we can set: $f(x) = 2x + 1$, which is the form we often use specifying a function.

- And in $f(x)$, we call x an *input variable*.

That's because it gets each of all the numbers in the domain, so each number is put *into* the variable x. And for the same reason, each number in the domain is called *an input value*, often just called *an input*, for short.

And for instance, putting an input 3 into x in the function f, we do it this way: $f(3)$.

Then, the function f produces the output this way: $f(3) = 2 \cdot 3 + 1 = 6 + 1 = 7$.

And naming (or specifying) a function more specifically, we can use another variable that can get the value of the function.   What value?

It is each and every number the function produces, and thus, is each and every output.

So for instance, if using y as the variable that gets the value of the function $f(x)$, we can name the function f this way, too: $y = f(x)$, and can specify the function f the way as follows: $y = f(x) = 2x + 1$, which is more specific than this: $f(x) = 2x + 1$.

- And we call y an *output variable*.

That's because it gets all the numbers (values) the function produces, and those numbers look getting *out* of the function.

And for the same reason, each of those numbers is called *an output value*, often just called *an output*, for short. And of course, the output variable gets one output at a time, and the output belongs to the range.

So what is a function?

A function is a math expression connecting two number sets called a domain and a range.

Connecting the two sets, the function generates a sequence of pairs of numbers as (0, 1), (3, 7), (8, 17), etc.

And for instance, in a pair (8, 17), 8 is called an input, and 17 is called the output for the input 8.   In short, in (8, 17), 8 is an input, and 17 is an output.

And putting an input 8 into x in the function $y = f(x) = 2x + 1$, we do it this way: $f(8)$.

Then, the function f produces the output this way: $f(8) = 2 \cdot 8 + 1 = 16 + 1 = 17$.
So in this case, we can call 17 the output for the input 8.

And the same is true for $f(8)$, too, since we have: $f(8) = 17$.
So we can call $f(8)$ the output for the input 8, too.

And we often say that the value of the function f is 17 when $x = 8$, or that the value of the function f is 17 for $x = 8$.

And we have: $y = f(x)$.

So the output variable y is said to get the value of f for each value of x.

And thus, we can say that $y = 17$ when $x = 8$, that $y = 17$ if $x = 8$, or that $y = 17$ for $x = 8$.

Practice 1 on Domains and Ranges

These are for familiarity with functions, particularly, domains and ranges. Some of these can be however, too easy to some students. Try them all, though.

0. Assuming $y = f(x) = x$ for x real, find the output for each input as follows.

0. $x = 0$.   1. $x = 1$.   2. $x = 3$.   3. $x = 0.2$.   4. $x = -0.7$.

1. Assuming $y = f(x) = 2x + 3$ for x real, find the output for each input as follows.

0. $x = 0$.   1. $x = 1$.   2. $x = 2$.   3. $x = 3$.   4. $x = -0.7$.   5. $x = -0.6$.   6. $x = -0.5$.

2. Assuming $y = f(x) = x^2 - 1$ for $-1 \leq x \leq 2$, find the output for each input as follows.

0. $x = 0$.   1. $x = 1$.   2. $x = -1$.   3. $x = 2$.   4. $x = -0.2$.   5. $x = 3$.   6. $x = -3$.

7. $x = -0.5$.   8. $x = 0.5$.   9. $x = -0.6$.   A. $x = -1.5$.   B. $x = 1.5$.

8

Suggestions or Solutions

To the **Problems** in the Example **0**

Assuming $y = f(x) = x$ for x real, find the output for each input as follows.

0. $x = 0$.   1. $x = 1$.   2. $x = 3$.   3. $x = 0.2$.   4. $x = -0.7$.

To begin with, what is the domain of the function f?

The function f is defined for all real numbers.
So 'for x real' means that every real number can be an input of the function f.
Thus, the domain of f is a set of all real numbers.

0. $x = 0 \Rightarrow y = f(0) = 0$, which is the output for the input **0**.

1. $x = 1 \Rightarrow y = f(1) = 1$, which is the output for $x = 1$.

2. $x = 3 \Rightarrow y = f(3) = 3$.

3. $x = 0.2 \Rightarrow y = f(0.2) = 0.2$.

4. $x = -0.7 \Rightarrow y = f(-0.7) = -0.7$.

So for each and every input, the output is the same as the input. In other words, every time the function f produces an output, the output is the same as the input. That is, in every pair of the input and output, both are *identical*.

So such a function is called an *identity function*.
And each of all the functions below is an identity function, too. So each is no other than each of all the others, and the same functions can have different names.

$y = f(x) = x$ for x real. $t = g(s) = s$ for s real. $v = h(u) = u$ for u real.

$q = k(p) = p$ for p real. $n = r(m) = m$ for m real.

Suggestions or Solutions
To the **Problems** in the Example **1**

Assuming $y = f(x) = 2x + 3$ for x real, find the output for each input as follows.

0. $x = 0$.   1. $x = 1$.   2. $x = 2$.   3. $x = 3$.   4. $x = -0.7$.   5. $x = -0.6$.   6. $x = -0.5$.

0. $x = 0 \Rightarrow y = f(0) = 2 \cdot 0 + 3 = 2 \times 0 = 0 + 3 = 3$, which is the output for $x = 0$.

1. $x = 1 \Rightarrow y = f(1) = 2 \cdot 1 + 3 = 2 + 3 = 5$, which is the output for the input **1**.

2. $x = 2 \Rightarrow y = f(2) = 2 \cdot 2 + 3 = 4 + 3 = 7$.

3. $x = 3 \Rightarrow y = f(3) = 2 \cdot 3 + 3 = 6 + 3 = 9$.

4. $x = -0.7 \Rightarrow y = f(-0.7) = 2 \cdot (-0.7) + 3 = -1.4 + 3 = 1.6$.

5. $x = -0.6 \Rightarrow y = f(-0.6) = 2 \cdot (-0.6) + 3 = -1.2 + 3 = 1.8$.

6. $x = -0.5 \Rightarrow y = f(-0.5) = 2 \cdot (-0.5) + 3 = -1 + 3 = 2$.

10

Suggestions or Solutions

To the **Problems** in the Example 2

Assuming $y = f(x) = x^2 - 1$ for $-1 \leq x \leq 2$, find the output for each input as follows.

0. $x = 0$.　　1. $x = 1$.　　2. $x = -1$.　　3. $x = 2$.　　4. $x = -0.2$.　　5. $x = 3$.　　6. $x = -3$.

7. $x = -0.5$.　　8. $x = 0.5$.　　9. $x = -0.6$.　　A. $x = -1.5$.　　B. $x = 1.5$.

0.　$x = 0 \Rightarrow y = f(0) = 0^2 - 1 = -1$, which is the output for $x = 0$.

1.　$x = 1 \Rightarrow y = f(1) = 1^2 - 1 = 0$, which is the output for the input **1**.

2.　$x = -1 \Rightarrow y = f(1) = (-1)^2 - 1 = 0$.

3.　$x = 2 \Rightarrow y = f(2) = 2^2 - 1 = 3$.

4.　$x = -0.2 \Rightarrow y = f(-0.2) = (-0.2)^2 - 1 = 0.04 - 1 = 0.96$.

5.　**3** is not in the domain, so the function f does not hold for $x = 3$.
So the output for $x = 3$, that is, $f(3)$ does not exist. That is, f is not defined for $x = 3$.

6.　**-3** is not in the domain, so the function f cannot be defined for $x = -3$.
In other words, -3 cannot be an input, so $f(-3)$ does not exist.

7.　$x = -0.5 \Rightarrow y = f(-0.5) = (-0.5)^2 - 1 = 0.25 - 1 = -0.75$.

8.　$x = 0.5 \Rightarrow y = f(0.5) = 0.5^2 - 1 = 0.25 - 1 = -0.75$.

9.　$x = -0.6 \Rightarrow y = f(-0.6) = (-0.6)^2 - 1 = 0.36 - 1 = -0.64$.

A.　**-1.5** is not included in the domain, so the function f cannot be defined for $x = -1.5$.
Therefore, $f(-1.5)$ does not exist.

B.　$x = 1.5 \Rightarrow y = f(1.5) = 1.5^2 - 1 = 2.25 - 1 = 1.25$.

0.1. How Functions Work 1

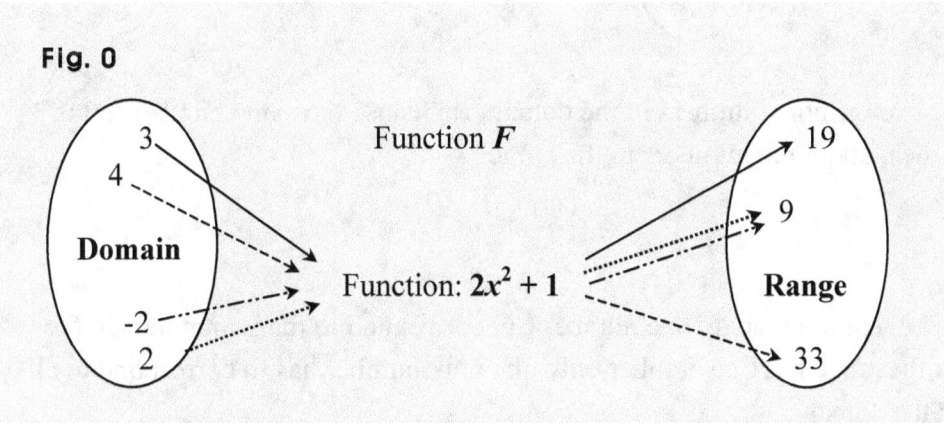

Fig. 0

So what is a function?

A function is an expression acting on a number set called a domain, and producing another set of numbers called a range.

- In short, a function is an expression that makes a range using a domain.

And we can notice that in the example above, some number in the range can correspond to more than one number in the domain. One number at a time though.

- Each number in a range corresponds to *at least one* number in the domain.

Therefore, for some functions, one number in the range can correspond to many numbers or even all the numbers in the domain. So for instance, we can have:

12

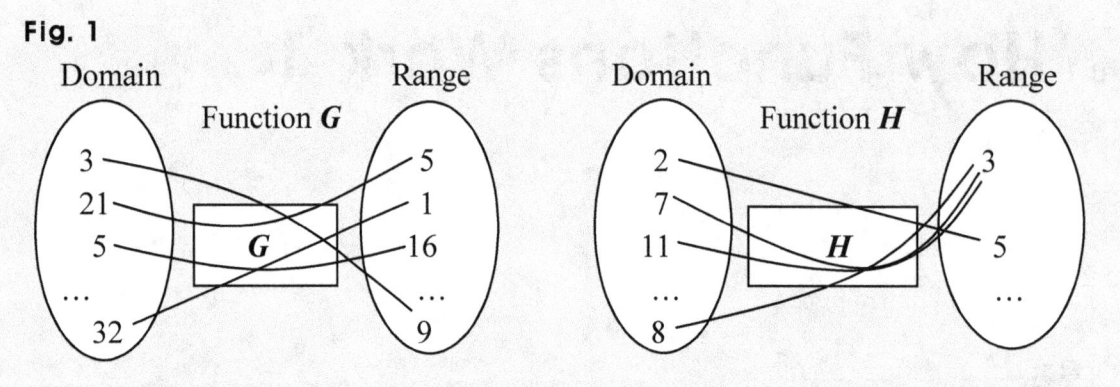

Fig. 1

Domain   Range   Domain   Range
Function *G*   Function *H*

That is because two or more numbers in the domain can cause the same number in the range. What then, about the numbers in the range?

As mentioned above, in the range, one number can correspond to many numbers in the domain. And if the range has one number only, the only number has to correspond to all the numbers in the domain.

Note however, for any function, two or more numbers in a range cannot correspond to the same number in a domain.

In other words, <u>no number</u> in the domain <u>can cause two or more</u> numbers in the range.

So no matter what function it may be, it has to produce one number only at a time.

For instance, we do **not** have a function as below. If it were a function, the two numbers 9 and 2 in the range wouldn't be able to correspond to the number 3 in the domain.

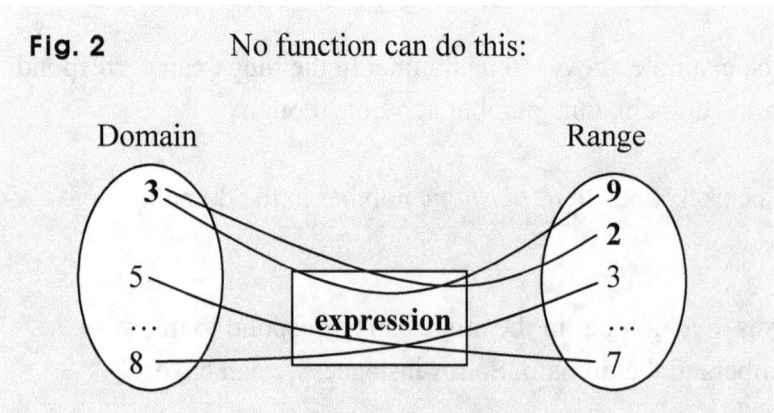

Fig. 2   No function can do this:

Domain   Range

Let's now, get back to the mechanical relationship.

We know that a machine basically repeats the same work.
A function is an expression made of math operations.

Every time a number is put into a function, the same group of operations repeats, and a number is produced.
So a function can be taken for a machine that performs a task over and over.

The task is a group of math operations as the ones in $x^2 - x + 1$.

What are the operations though, in $x^2 - x + 1$?

We can see three:

One is: $x^2 = x \cdot x$, which is a multiplication.
Another is: $x^2 - x$, which is a subtraction.
And the other is: $(x^2 - x) + 1$, which is an addition.

Doing this: x^2, we get a product. And using the product, we do this: $x^2 - x$.
Then, using the difference, we do this: $(x^2 - x) + 1$. And then, we get the sum.
The sum is a number that the expression $x^2 - x + 1$ produces.
That is, it's an output that the function $x^2 - x + 1$ produces for an input.

And every time the set of operations gets completed, an output gets produced.
So for instance:

If 1 gets into $x^2 - x + 1$, we get: $1^2 - 1 + 1 = 1$.

If 2 gets into $x^2 - x + 1$, we get: $2^2 - 2 + 1 = 3$.

If 3 gets into $x^2 - x + 1$, we get: $3^2 - 3 + 1 = 7$.
...

And we can get put it this way, too:

$$1 \longrightarrow x^2 - x + 1 \longrightarrow 1$$

So we get: (1, 1), where 1 is an input, and 1 is the output.

$$2 \longrightarrow x^2 - x + 1 \longrightarrow 3$$

So we get: (2, 3), where 1 is from the domain, and 3 belongs to the range.

$$3 \longrightarrow x^2 - x + 1 \longrightarrow 7. \quad \text{So we get: (3, 7).}$$

And assuming X is the domain, and Y is the range, and putting at once all the function executions for all the inputs in X, we can put it the way below:

$$X \longrightarrow x^2 - x + 1 \longrightarrow Y$$

And assuming F is the function $x^2 - x + 1$, we can simplify the idea the way below:

$$X \xrightarrow{F} Y.$$

Next, depending on the nature of the correspondence a function creates, the type of the function can get determined. One is called *one-to-one*, and another is called *many-to-one*.

What then, do we mean by one-to-one?

Suppose no two numbers in the domain cause the same number in the range.

That is, every number in the domain causes a different number in the range. In other words, every output is different.

Then, the function is said to be one-to-one. Otherwise, it is many-to-one.

What then, about one-to-many?

There is no such a function. That's because a function produces one number only at a time. In other words, in any function, no two numbers can be caused by a number.

So a function is either one-to-one or many-to-one.
And we can put the idea the way below, too:

If all the numbers in the range are used once only, it is one-to-one.
So every output is different. And thus, if $f(a) \neq f(b)$ where $a \neq b$, f is one-to-one.

If some or all the numbers in the range are used more than once, it is many-to-one.
So some or all the outputs can be the same. For instance:

If $f(2) = f(3) = 1$, f is many-to-one.
And also, if $g(x) = 2$ for all the values of x, g is many-to-one, too.

And as mentioned earlier, in high school math and calculus, all the numbers in both the domain and the range have to get connected. And such a function is said to be *onto*, and thus, is called an onto function.

So more specifically, a function is either one-to-one and onto or many-to-one and onto.

Working with a function then, do we need to see if a function is one-to-one or not?

Sometimes, we need to come up with a new function from a particular function. Coming up with the function, we have to use as the domain the range of the function particular, and use as the range the domain of the particular function. And that's not it.

We need to maintain the existing correspondence between numbers in both sets, the domain and the range of the particular function.

For instance, if the particular function makes (1, 2), (4, 7), (6, -5) etc., the new function has to make (2, 1), (7, 4), (-5, 6), etc.

We call such a new function *an inverse function*.

And we call the new function stated above, the inverse of the particular function, and often just call it *the inverse*, for short.

Then, the function particular has to be one-to-one. How come?

Taking the inverse of a function many-to-one, we cannot get a function.
Or rather, we cannot take the inverse of a function many-to-one. Why not though?

If a function is many-to-one, the inverse would be one-to-many, and thus, would not be a function. A function cannot be one-to-many.

  • Each number in a domain gets connected to one number only in the range.

So no two numbers in the domain can be connected to one same number in the range.

So what is a function?

A function is an expression that connects two number sets, called a domain and a range in a way that every time the function gets executed, a *pair* of numbers gets determined.

And number pairs get determined in either of the two ways below:

(2, 1), **(4, 3)**, **(7, 5)**,etc.   Then, the function f is one-to-one, since $f(a) \neq f(b)$ if $a \neq b$.

(1, 3), **(5, 9)**, **(7, 3)**, etc.   Then, the function f is many-to-one, since $f(1) = f(7)$.

The first one in each pair is from the domain, and the other belongs to the range.

And that's the way the two sets, that is, the domain and the range get connected.

But no function can generate number pairs the way as follows: **(1, 2)**, **(1, 9)**, **(3, 4)**, etc.

Practice 2 on Domains and Ranges

These are too, for familiarity with functions, particularly, domains and ranges. And these are for your algebra skill, too. And the algebra is on inequalities, which are sometimes not quite straightforward. So let's find now, the range of each function as follows.

0. $y = f(x) = x$ for x real.

1. $y = f(x) = x$ for $0 \leq x \leq 1$.

2. $y = f(x) = 2x$ for x real.

3. $y = f(x) = 2x$ for $0 \leq x \leq 1$.

4. $y = f(x) = 2x - 3$ for $0 \leq x \leq 1$.

5. $y = f(x) = x^2$ for x real.

6. $y = f(x) = x^2$ for $x \geq 0$.

7. $y = f(x) = x^2$ for $x \leq 0$.

8. $y = f(x) = x^2$ for $0 \leq x \leq 1$.

9. $y = f(x) = x^2$ for $x \geq 1$.

A. $y = f(x) = x^2$ for $x \geq -1$.

B. $y = f(x) = x^2$ for $x \leq 1$.

C. $y = f(x) = x^2$ for $1 \leq x < 2$.

D. $y = f(x) = x^2$ for $1 \leq x \leq 2$ or $-2 \leq x \leq -1$.

Suggestions or Solutions
To the Problems in Practice 2

0. Find the range of a function where $y = f(x) = x$ for x real.

First, what is the domain of a function?

The domain is the set of all the inputs, which are all the numbers that the input variable can get, so a function can hold (be defined) for only the numbers in the domain.

Next, we have: $y = f(x) = x$ for x real.

And 'for x real' means that every real number can be an input of the function.
That is, the domain is a set of all real numbers.

So in the function f, x is the input variable, and can get every real number.
Therefore, the function f can hold for all real numbers.

Next, *the range is the set of all the outputs*. And the function f is an identity function.

So after every execution of the function f, the output is the same as the input.
That is, whenever f produces an output, the output is the same as the input.

So the range is the same as the domain, and thus, is a set of all real numbers, too.

And such a function as f is said to be *one-to-one*, too, so *all the outputs* are *different*.
What then, about a many-to-one function?

By a function many-to-one, at least one output gets produced more than once.
So if a function is one-to-one, it produces at least two identical outputs.
That is, some outputs have to be the same.
And as an extreme example, we can have a constant function.   What then, is constant?

Every output is the same, so the output is constant, and thus, the value of the function is constant. In such a function, the domain has many numbers, that is, many inputs, yet the range has only one number, that is, every output is always the same.

So for instance, we can have: $y = g(x) = 3$, and every output is 3 for all the inputs.

Thus, we get: $g(1) = g(-5) = g(7) = g(0.3) = \dots = 3$.

1. **Find the range of a function $y = f(x) = x$ for $0 \leq x \leq 1$.**

In this case, too, the range is the same as the domain. How come?

The function f is an identity function, so after each and every execution of f, the output is the same as the input. So the range is the same as the domain.
What is the domain though, of the function f?

The domain is the set of all the inputs. And the input variable gets every input.
And the input variable is x, and we have this: $0 \leq x \leq 1$.
So the domain is the set of all the real numbers from 0 to 1.
Specifying thus, the domain of f, we can put it this way: $0 \leq x \leq 1$.
And we say that the function f is defined for all the real numbers from 0 to 1.

Next, the range is the same as the domain, so the range, too, is a set of all the real numbers from 0 to 1.
And the range is the set of all the outputs. Of a function then, what gets every output?

The output variable does. And of the function $y = f(x)$, the output variable is y.
So y gets each and every output, that is, all the numbers in the range.
And we know the range is a set of all the real numbers from 0 to 1.
So specifying the range, we can put it this way: $0 \leq y \leq 1$.

And thus, specifying the range, we often do it specifying the extent of the input variable, and, specifying the domain, too, we do it specifying the extent of the output variable.

2.   Find the range of a function $y = f(x) = 2x$ for x real.

First, that x is real means that every real number can be an input of the function.
So the domain is a set of all real numbers.
Thus, the function f can hold for all real numbers.

Next, the expression of the function f is: $2x$.
That is, after every execution of f, the output is twice the input, which is a real number.
So each of all the outputs is twice each of all the inputs, and thus, is a real number.

So every real number can be an output.
So the set of all the outputs is a set of all real numbers, too.
Therefore, the range is a set of all real numbers, too.

So is the same true, too, for any function whose domain is a set of all real numbers?

In other words, if a function can hold for all real numbers, is the range a set of all real numbers, too?

Not necessarily, and we will see why not shortly.

3.   Find the range of a function $y = f(x) = 2x$ for $0 \leq x \leq 1$.

First, the domain is: $0 \leq x \leq 1$, so the function f is defined for $0 \leq x \leq 1$.

And next, the expression of f is: $2x$. So each output is twice each input.

The output variable y gets each output, and we have: $y = 2x$, where x gets each input.

And we have: $0 \leq x \leq 1$, too, so we get: $0 \leq x \leq 1 \Rightarrow 0 \leq 2x \leq 2 \Rightarrow 0 \leq y \leq 2$.

We know the range is the set of all the outputs, and y gets each and every output.

So the range is: $0 \leq y \leq 2$.

4.   **Find the range of a function $y = f(x) = 2x - 3$ for $0 \leq x \leq 1$.**

To begin with, what is the domain of the function f?

The domain is: $0 \leq x \leq 1$, So the function f is defined for all the real numbers from 0 to 1.

In the expression $2x - 3$, each input is multiplied by 2, and 3 is subtracted from the product, then the difference is the output, which is put into y, the output variable.

And we have: $y = 2x - 3$.

So we get: $0 \leq x \leq 1 \Rightarrow 0 \leq 2x \leq 2 \Rightarrow 0 - 3 \leq 2x - 3 \leq 2 - 3 \Rightarrow -3 \leq y \leq -1$.

And the range is the set of all the outputs, that is, the set of all the y-values.

Therefore, the range is: $-3 \leq y \leq -1$.

5.   **Find the range of a function $y = f(x) = x^2$ for x real.**

The domain is a set of all real numbers, so f can hold for all real numbers.
And the expression is: x^2. So each output is the square of each input.   So?

No matter what a real number may be, the square of it is ≥ 0.   So?

Each output is ≥ 0. And y gets each output. Thus, the range is: $y \geq 0$.

6.   **Find the range of a function $y = f(x) = x^2$ for $x \geq 0$.**

First, the domain is: $x \geq 0$, so the function f can hold for all the real numbers ≥ 0.

Next, the expression is: x^2, and the square of a real number ≥ 0 is positive or 0, too. So?

22

Each output is ≥ 0. And y gets each output. Thus, the range is: $y \geq 0$.

7. Find the range of a function $y = f(x) = x^2$ for $x \leq 0$.

First, the domain is: $x \leq 0$, so the function f is defined for all the real numbers ≤ 0.
Next, the expression is: x^2, and the square of a real number ≤ 0 is ≥ 0. How come?

The square of any number negative is positive, and $0^2 = 0$.
So every output is ≥ 0. And y gets each output. Thus, the range is: $y \geq 0$.

8. Find the range of a function $y = f(x) = x^2$ for $0 \leq x \leq 1$.

First, the domain is: $0 \leq x \leq 1$, so the function f holds for all the real numbers from 0 to 1.

Next, we know the range is the set of all the outputs, and the output variable gets each and every output. So the range is the set of all the numbers the output variable gets.

The output variable is y, and we have: $y = x^2$.

And we have: $0 \leq x \leq 1$, too.

So we get: $0 \leq x \leq 1 \Rightarrow 0 \leq x^2 \leq 1 \Rightarrow 0 \leq y \leq 1$. Thus, the range is: $0 \leq y \leq 1$.

9. Find the range of a function $y = f(x) = x^2$ for $x \geq 1$.

The domain is: $x \geq 1$, so the function f is defined for $x \geq 1$. And the expression is: x^2.

So we get: $x \geq 1 \Rightarrow x^2 \geq 1 \Rightarrow y \geq 1$, since $y = f(x) = x^2$. And thus, the range is: $y \geq 1$.

A.   Find the range of a function $y = f(x) = x^2$ for $x \geq$ -1.

The domain is: $x \geq$ **-1**, so the function f is defined for $x \geq$ **-1**. And the expression is: x^2.

So is the square of such a real number greater than or equal to 1?
In other words, is it the case where: $x \geq$ **-1** $\Rightarrow x^2 \geq$ **1**?

No, it is not the case. The square of a real number $\geq$ **-1** is $\geq$ **0**.   How come?

First, the square of a real number is ≥ 0, no matter what real number it may be.

Next, in f, x gets each input, which can be 0, and the square of 0 is 0. So x^2 can be 0.

So it the case where: $x \geq$ **-1** $\Rightarrow x^2 \geq$ **0**.   Therefore, the range is: $y \geq$ **0**.

It is in fact, <u>not a good idea</u> to apply inequalities to find the range if the function is not linear as **2x** or **5x – 3**, etc. What then, is a good idea?

Usually, using the graph of a function, we can get the range readily and quickly.
And more importantly, it is safe to get the range graphically.
So it's not only easy and fast but safe, too, to find the range using a graph.

So graphing matters.

The examples though, in this practice, are quite straightforward even though they are not linear. So you can find the ranges without much difficulty even if not using the graphs.

And it's in fact, a good idea to do some examples without using graphs, too, if the examples are not quite complicated, of course. And doing those examples, you can see better how algebra works with inequalities, so you get to improve your algebra, of course, and get used to tracking outputs changing as inputs change. So you are going to do some more examples with no graphs used in this practice set and in the next set, too.

24

B. Find the range of a function $y = f(x) = x^2$ for $x \leq 1$.

The domain is: $x \leq 1$, so the function f is defined for $x \leq 1$. And the expression is: x^2.

The square of a real number ≤ 1 is greater than or equal to not 1 but 0.   How come?

First, the square of a real number is ≥ 0.

Next, in f, x gets each input, which can be 0, and the square of 0 is 0. So x^2 can be 0.

So it the case where: $x \leq 1 \Rightarrow x^2 \geq 0$. Therefore, the range is: $y \geq 0$.

C. Find the range of a function $y = f(x) = x^2$ for $1 \leq x < 2$.

The domain is: $1 \leq x < 2$, and we need to note that the domain does not include 2. So the function f is defined for $1 \leq x < 2$, and not for $x = 2$.

The output variable is y, and we have: $y = x^2$.

So we get: $1 \leq x < 2 \Rightarrow 1 \leq x^2 < 4 \Rightarrow 1 \leq y < 4$. Therefore, the range is: $1 \leq y < 4$.

D. Find the range of a function $y = f(x) = x^2$ for $1 \leq x \leq 2$ or $-2 \leq x \leq -1$.

The domain is: $1 \leq x \leq 2$ or $-2 \leq x \leq -1$. So f is defined for $1 \leq x \leq 2$ or $-2 \leq x \leq -1$.
The output variable is y, and we have: $y = x^2$. So we get:

$1 \leq x \leq 2 \Rightarrow 1 \leq x^2 \leq 4 \Rightarrow 1 \leq y \leq 4$.

$-2 \leq x \leq -1 \Rightarrow 1 \leq x^2 \leq 4 \Rightarrow 1 \leq y \leq 4$, which happens to be the same as the one above.

Thus, we get: $1 \leq y \leq 4$, which is the range.

0.2. How Functions Work 2

Why is $f(x)$ though, read as f of x?

Saying f of x, we are saying that f (is a function) of x. So in short, we just say: f of x.

And saying f is a function of x, we mean $f(x)$ changes as x changes.

In short, f varies as x varies.

More specifically, as the value of x changes, the value of $f(x)$ changes.

Suppose for instance, $y = f(x) = 2x + 5$. Then:

If $x = 1$, we get: $y = f(1) = 2 \cdot 1 + 5 = 7$, so we get: $y = f(1) = 7$. That is, $y = 7$ if $x = 1$.

If $x = 2$, we get: $y = f(2) = 2 \cdot 2 + 5 = 9$, so we get: $y = f(2) = 9$. That is, $y = 9$ if $x = 2$.

So as the value of x changes, the value of $f(x)$ changes, and so does the value of y, of course. And we know x gets the inputs. And as the input changes, the output changes.

Saying thus, a function f of x, or just f of x, we are saying that f is a function of x, and that x is the input variable.

And we mean that the value of f or the value of $f(x)$ changes as the value of x changes. Also of course, since $y = f(x)$, the values of y changes as the value of x changes.

Suppose for another instance, $t = g(s) = 3s + 4$. Then:

If $s = 1$, we get: $t = g(1) = 3 \cdot 1 + 4 = 7$, so we get: $t = g(1) = 5$. That is, $t = 7$ if $s = 1$.

If $s = 2$, we get: $t = g(2) = 3 \cdot 2 + 4 = 10$, so we get: $t = g(2) = 10$. That is, $t = 10$ if $s = 2$.

So as s changes, the value of $g(s)$ changes, and so does the value of t, of course.
And thus, s gets the inputs, $g(s)$ gets the outputs, and so does t, since $t = g(s)$.

Saying thus, a function g of s, or just g of s, we are saying that g is a function of s, and that s is the input variable.

And we mean that the value of g or $g(s)$ changes as the value of s changes.
And of course, since $t = g(s)$, the value of t changes as the value of s changes, too.

And for another instance, saying <u>a function v equals h of u</u>, we mean: $v = h(u)$, and we are saying that h is a function of u, and u is the input variable.

And we mean that the value of h or the value of $h(u)$ changes as the value of u changes.

  • Specifying however, a function the way above, that is, with no domain specified, we just assume the largest possible domain.

Given for instance, a function this way: $y = f(x) = 2x + 5$, we are to assume that the domain is a set of all real numbers, because it is the largest possible domain.

How come it is the largest?

Every real number can be an input of the function f. In other words, the function f can be defined for every real number. So the largest domain is a set of all real numbers.

  • So just setting: $y = f(x) = 2x + 5$, we mean: $y = f(x) = 2x + 5$ for x real.

What if the domain of the function f above is a set of all numbers grater than 3?

Then, we can put it this way: $y = f(x) = 2x + 5$ for $x > 3$.

What then, is the range of the function f above?

We have: $x > 5$. So we can get: $x > 5 \Rightarrow 2x > 10 \Rightarrow 2x + 1 > 11$.

And we have: $y = f(x) = 2x + 1$ for $x > 5$. So we get: $y = f(x) > 11 \Rightarrow y > 11$.

And we know y is the output variable, so it gets all the outputs that belong to the range. The range is thus, a set of all numbers bigger than 11, and can be put this way: $y > 11$.

And normally, if asked to find a domain, we are to find the largest possible domain. What then, is the domain if the function given is: $y = g(x) = \sqrt{x+1}$?

- We know what's inside the square root sign is positive or 0.

So we get: $x + 1 \geq 0 \Rightarrow x \geq -1$, and thus, the (largest possible) domain is: $x \geq -1$.

If any number less than -1 is put into (x in) the function g, we don't get a real number.

So for instance, we cannot make a function as $y = g(x) = \sqrt{x-2}$ for $x > 1$.

That's because if $x = 1.5$, we get: $g(1.5) = \sqrt{1.5 - 2} = \sqrt{-0.5}$, which is not real.

What kind of functions then, do we work with in high school math?

Based on expressions that functions have, we can classify or categorize functions:

An expression as $x^2 - x + 1$ is said to be algebraic, so if a function has an algebraic expression, the function is called an algebraic function. And if a function is not an expression algebraic, the function is called a transcendental function.

For instance, expressions as $\sin x$, 3^x, $\log x$, $\sqrt{x+1}$, and $\frac{x}{x^2+1}$ are not algebraic, so if a function is such an expression, the function is said to be transcendental. In particular:

A function $y = f(x) = \sin x$ for $x \geq \pi$ is called a trigonometric function, and more specifically, it is called a sine function.

A function $y = g(x) = 3^x$ for $x > -2$ is called an exponential function.

A function $y = h(x) = \log x$ for $x > 0$ is called a log function.

A function $y = p(x) = \sqrt{x+1}$ for $x \geq 2$ is called a radical (or irrational) function.

A function $y = q(x) = \frac{x}{x^2+1}$ for $x \geq 1$ is called a fractional (or rational) function.

So in high school math, we work with some of those transcendental functions, together with algebraic ones.

And moving on to calculus in a college or university, we get to work with a function more advanced, of which the domain can have two or more sets of numbers.

For instance, we get to work with functions as $z = r(x, y) = x + y + 1$ for $x > 0$, and $y < 1$.

The function r above has not one but two input variables, and the two are x and y.
So the domain is not just one set, but is made of two sets:

• One is for the input variable x, and is a set of all numbers greater than 0, that is, a set of all positive numbers.

• And the other is for the other input variable y, and is a set of all numbers less than 1.

And of course, z is the output variable of the function r.
Every function has one output variable. So every range is just one set of numbers.

Of a function basic, the domain is made of one set of numbers.

And such a basic function can be called a 2-D function, because it has two data sets, since it has two variables, one is an input variable, and the other is an output variable.

So if the domain is made of two sets, the function can be said to be 3-D, because it has three data sets, since it has three variables, two are input variables, and the other is an output variable.

And if the domain is composed of three sets, the function can be called a 4-D function, because it has four data sets, since it has four variables, three are input variables, and the other is an output variable.

And so forth.

So a function has quite a few: a name, an expression, a domain, and a range.
That's not it though.

A function has its curve, too.
So we can put a function in a place called a graph. Putting thus, a function in a graph, we put the curve in a graph. Where do we put the graph, though?

We put a graph in a coordinate system.
If a function has two variables, an input variable and an output variable, we can say that we put the function in a coordinate system said to be 2-D. And we just call it a 2-D system, for short.
In a 2-D system, we usually put two axes perpendicular to each other, which means therefore, a plane.
That's because we can use one axis for a length, and use the other for a width.
So usually, we just call such a 2-D system a coordinate plane.
What then, about 3-D system?

We can call it a coordinate space, because in such a system, we put three axes perpendicular to each other. One is for a length, another is for a width, and the other is for a height. (Details on graphs and curves are covered in **GRAPH OPERATIONS**.)

And we can name a coordinate system, too, and can do so using the variables used in the function to be put in the system.

So for instance, assuming we use *x* as the input variable, and use *y* as the output variable in a function *F*, and putting the curve of *F* in a graph, we put the graph in the *x-y* coordinate system, often called briefly the *x-y* system, too, which is made of the *x*-axis and the *y*-axis perpendicular to each other, and thus, can be said to form a plane.

So we often just call the *x-y* system the *x-y* plane, too.
In a function put the *x-y* system, *x* is the input variable, and *y* is the output variable.
For instance, putting the function *F* in the *x-y* system, we can set: $y = F(x)$.
And we can say that we can put the curve of *F* in the *x-y* plane or the *x-y* system.
Why curve though?

Putting the curve of a function in a graph, we can actually see the function, and more importantly, we can see how the function behaves. So solving problems with functions, *we can get the solutions easier and quicker* putting the functions in their graphs.

(Details on graphs and curves are covered in **GRAPH OPERATIONS**.)

So a function is an expression acting on a number set called the domain, and producing another set called the range, and can actually show itself if it's put in a graph.

Can we have though, a function that is just a number as 1?
That is, can we have a function as $y = f(x) = 1$?

Yes, we can. Technically, a number itself can be an expression, too, since it expresses a value. And from a function's point of view, we can put the idea the way below:

As a range, we can use a set of one number only as {1} or {0}. How come?

We can have such a function, and it is called a many-to-one function. So if the range has one number only, the only number corresponds to all the numbers in the domain.
  • And we often call it a *constant function.*

In such a case, though the input changes, the output does not change. So in case of a constant function, the output is constant, and thus, the curve is a line parallel to the axis where all the inputs are placed. So for instance, being in the *x-y* plane, the line is parallel to the *x*-axis. And for instance, the functions below are constant functions:

$y = f(x) = 1$ for $x \geq 1$.   The value of $f(x)$ is 1 if x is bigger than or equal to 1.

So for instance, we can get: $f(1) = f(1.01) = f(2) = f(21) = 1$.

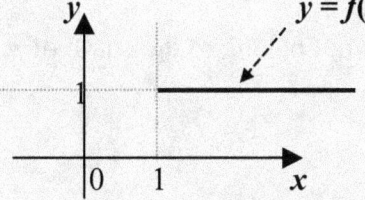

Fig. 0 $y = f(x)$ The curve of f is a ray.

$t = g(s) = 9$ for s real.   The value of $g(s)$ is 9 if s is a real number.

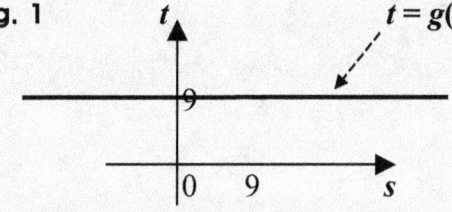

Fig. 1 $t = g(s)$ The curve of g is a line.

$u = q(v) = k$ for $v > 2$, where k is a constant > 0.

The value of $q(v)$ is k if v is bigger than 2.

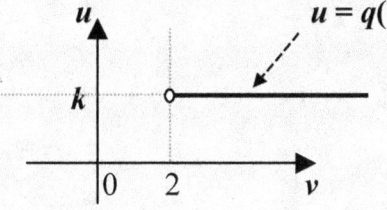

Fig. 2 $u = q(v)$ The curve of q is a ray.

Note that the point $(2, k)$ does not belong to the ray, which is the curve of q, of course. And of course, if $k < 0$, the ray will be below the *x*-axis.

32

$y = p(x) = 2$ for $2 < x < 6$.   The value of $p(x)$ is 2 if x is between 2 and 6.

Fig. 3 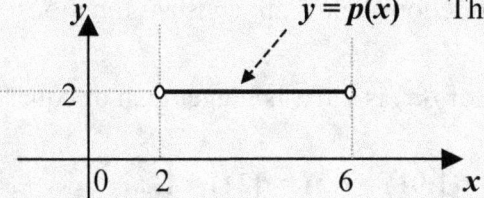 $y = p(x)$ The curve of p is a line segment.

Note that the two points **(2, 2)** and **(6, 2)** do not belong to the curve of p.

$y = h(x) = 0$.   In particular, we call such a function as h a zero function.

And of course, the curve of the zero function h is the x-axis itself.

Fig. 4 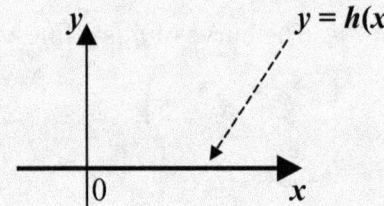 $y = h(x)$

Practice 3 on Domains and Ranges

These are also, for familiarity with functions, particularly, domains and ranges. And also, these are for your algebra on inequalities, which are sometimes not quite straightforward.

Find the range of each function as follows.

0. $y = f(x) = x^2$ for $0 \leq x < 1$ or $-2 \leq x \leq -1$.

1. $y = f(x) = x^2$ for $0 \leq x < 0.1$ or $-2 \leq x < -1$.

2. $y = f(x) = -x^2$ for x real.

3. $y = f(x) = -x^2$ for $x \geq 0$.

4. $y = f(x) = -x^2$ for $x \leq 0$.

5. $y = f(x) = -x^2$ for $0 \leq x \leq 1$.

6. $y = f(x) = -x^2$ for $x \geq 1$.

7. $y = f(x) = -x^2$ for $x \geq -1$.

8. $y = f(x) = -x^2$ for $x \leq 1$.

9. $y = f(x) = -x^2$ for $1 \leq x < 2$.

34

A. $y = f(x) = -x^2$ for $1 \le x \le 2$ or $-2 \le x \le -1$.

B. $y = f(x) = -x^2$ for $0 \le x < 1$ or $-2 \le x \le -1$.

C. $y = f(x) = -x^2$ for $0 \le x < 0.1$ or $-2 \le x < -1$.

D. $y = f(x) = 2x^2$ for x real.

E. $y = f(x) = 2x^2$ for $x \ge 0$.

F. $y = f(x) = 2x^2$ for $x \le 0$.

G. $y = f(x) = 2x^2$ for $0 \le x \le 1$.

H. $y = f(x) = 2x^2$ for $x \ge 1$.

I. $y = f(x) = -2x^2$ for $x \ge -1$.

J. $y = f(x) = -2x^2$ for $x \le 1$.

K. $y = f(x) = -2x^2$ for $1 < x \le 2$.

L. $y = f(x) = 2x^2 + x$ for $1 \le x \le 2$ or $-2 \le x \le -1$.

Suggestions or Solutions
To the Problems in Practice 3

To begin with, when finding ranges of functions nonlinear as the ones here, we may not want to apply inequalities only.

Usually, using the graph of a function, you can get the range readily and quickly. And more importantly, you can get it safely.

The examples though, in this practice, are quite straightforward even though they are not linear. So you can find the ranges without much difficulty even if using no graphs. And it's in fact, a good idea to do some examples without using graphs, too, if the examples are not quite complicated, of course.

Doing most of these examples with no graphs, you can improve your algebra on inequalities, and improve your skill of tracking outputs changing as inputs change.

0.   **Find the range of a function $y = f(x) = x^2$ for $0 \le x < 1$ or $-2 \le x \le -1$.**

The domain is: $0 \le x < 1$ or $-2 \le x \le -1$. So f holds for $0 \le x < 1$ or $-2 \le x \le -1$.

The output variable is y, and we have: $y = x^2$. So we get:

$0 \le x < 1 \Rightarrow 0 \le x^2 < 1 \Rightarrow 0 \le y \le 1$.

$-2 \le x \le -1 \Rightarrow 1 \le x^2 \le 4 \Rightarrow 0 \le y \le 4$.

Thus, we get: $0 \le y \le 1$ or $0 \le y \le 4$, so we get: $0 \le y \le 4$, which is the range.

1.   **Find the range of a function $y = f(x) = x^2$ for $0 \le x < 0.1$ or $-2 \le x < -1$.**

The domain is: $0 \le x < 0.1$ or $-2 \le x < -1$. So f can hold for $0 \le x < 0.1$ or $-2 \le x < -1$.

The output variable is y, and we have: $y = x^2$. So we get:

$0 \le x < 0.1 \Rightarrow 0 \le x^2 < 0.01 \Rightarrow 0 \le y < 0.01$.

$-2 \le x < -1 \Rightarrow 1 < x^2 \le 4 \Rightarrow 1 < y \le 4$.

So we get: $0 \le y < 0.01$ or $1 < y \le 4$. And we know y gets each and every output of f.

Therefore, range is: $0 \le y < 0.01$ or $1 < y \le 4$.

2.   Find the range of a function where $y = f(x) = -x^2$ for x real.

That x is real means that any real number can be an input. That is, the domain is a set of all real numbers. So the function f holds for all real numbers.

- Whatever a real number it may be, the square of it is ≥ 0.

However, the expression is: $-x^2$, so each output is the negative of the square of each input. Thus, each output is less and or equal to 0.
And we know y gets each and every output of the function f. So the range is: $y \leq 0$.

3.   Find the range of a function $y = f(x) = -x^2$ for $x \geq 0$.

The domain is: $x \geq 0$. So the function f holds for $x \geq 0$. And the expression is: $-x^2$.
And we have: $y = -x^2$, too.

So we get: $x \geq 0 \Rightarrow x^2 \geq 0 \Rightarrow -x^2 \leq 0 \Rightarrow y \leq 0$. Therefore, the range is: $y \leq 0$.

4.   Find the range of a function $y = f(x) = -x^2$ for $x \leq 0$.

The domain is: $x \leq 0$. So the function f holds for $x \leq 0$. And the expression is: $-x^2$.
And we have: $y = -x^2$, too.

So we get: $x \leq 0 \Rightarrow x^2 \geq 0 \Rightarrow -x^2 \leq 0 \Rightarrow y \leq 0$. Therefore, the range is: $y \leq 0$.

5.   Find the range of a function $y = f(x) = -x^2$ for $0 \leq x \leq 1$.

The domain is: $0 \leq x \leq 1$. So the function f holds for $0 \leq x \leq 1$. And the expression is: $-x^2$.
And we have: $y = -x^2$, too.

So we get: $0 \leq x \leq 1 \Rightarrow 0 \leq x^2 \leq 1 \Rightarrow -1 \leq -x^2 \leq 0 \Rightarrow -1 \leq y \leq 0$, which is the range.

6. **Find the range of a function $y = f(x) = -x^2$ for $x \geq 1$.**

The domain is: $x \geq 1$. So the function f holds for $x \geq 1$. And the expression is: $-x^2$.
And we have: $y = -x^2$, too.
So we get: $x \geq 1 \Rightarrow x^2 \geq 1 \Rightarrow -x^2 \leq -1 \Rightarrow y \leq -1$, which is the range.

7. **Find the range of a function $y = f(x) = -x^2$ for $x \geq -1$.**

The domain is: $x \geq -1$. So the function f holds for $x \geq -1$. And the expression is: $-x^2$.

The square of a real number ≥ -1 is greater than or equal to not 1 but 0.

In f, x gets each input, which can be 0, and the square of 0 is 0.

However, the expression is: $-x^2$, so each output is the negative of each square.
Thus, each output is less than or equal to 0.

In other words, it the case where: $x \geq -1 \Rightarrow x^2 \geq 0 \Rightarrow -x^2 \leq 0$.

And we have: $y = -x^2$, too. Therefore, the range is: $y \leq 0$.

8. **Find the range of a function $y = f(x) = -x^2$ for $x \leq 1$.**

The domain is: $x \leq 1$. So f holds for $x \leq 1$. And the expression is: $-x^2$.

The square of a real number ≤ 1 is greater than or equal to not 1 but 0.
That's because such a real number can be 0, and the square of 0 is 0.

However, the expression is: $-x^2$, so each output is the negative of each square.
In other words, it the case where $x \leq 1 \Rightarrow x^2 \geq 0 \Rightarrow -x^2 \leq 0$.

And we have: $y = -x^2$, too. Therefore, the range is: $y \leq 0$.

38

9.   Find the range of a function $y = f(x) = -x^2$ for $1 \le x < 2$.

The domain is: $1 \le x < 2$, and we need to note that the domain does not include 2.

So f holds for $1 \le x < 2$, and not for $x = 2$.

The output variable is y, and we have: $y = -x^2$.

So we get: $1 \le x < 2 \Rightarrow 1 \le x^2 < 4 \Rightarrow -4 < -x^2 \le -1 \Rightarrow -4 < y \le -1$, which is the range.

A.   Find the range of a function $y = f(x) = -x^2$ for $1 \le x \le 2$ or $-2 \le x \le -1$.

The domain is: $1 \le x \le 2$ or $-2 \le x \le -1$. So f holds for $1 \le x \le 2$ or $-2 \le x \le -1$.

The output variable is y, and $y = -x^2$. So we get:

$1 \le x \le 2 \Rightarrow 1 \le x^2 \le 4 \Rightarrow -4 \le -x^2 \le -1 \Rightarrow -4 \le y \le -1$.

$-2 \le x \le -1 \Rightarrow 1 \le x^2 \le 4 \Rightarrow -4 \le -x^2 \le -1 \Rightarrow -4 \le y \le -1$, the same the one above.

Thus, we get: $-4 \le y \le -1$, which is the range.

B.   Find the range of a function $y = f(x) = -x^2$ for $0 \le x < 1$ or $-2 \le x \le -1$.

The domain is: $0 \le x < 1$ or $-2 \le x \le -1$. So f holds for $0 \le x < 1$ or $-2 \le x \le -1$.

The output variable is y, and $y = -x^2$. So we get:

$0 \le x < 1 \Rightarrow 0 \le x^2 < 1 \Rightarrow -1 \le -x^2 \le 0 \Rightarrow -1 \le y \le 0$.

$-2 \le x \le -1 \Rightarrow 1 \le x^2 \le 4 \Rightarrow -4 \le -x^2 \le 0 \Rightarrow -4 \le y \le 0$.

Thus, we get: $-1 \le y \le 0$ or $-4 \le y \le 0$, so we get: $-4 \le y \le 0$, which is the range.

C. Find the range of a function $y = f(x) = -x^2$ for $0 \leq x < 0.1$ or $-2 \leq x < -1$.

The domain is: $0 \leq x < 0.1$ or $-2 \leq x < -1$. So f can hold for $0 \leq x < 0.1$ or $-2 \leq x < -1$.

The output variable is y, and we have: $y = -x^2$. So we get:

$0 \leq x < 0.1 \Rightarrow 0 \leq x^2 < 0.01 \Rightarrow -0.01 < -x^2 \leq 0 \Rightarrow -0.01 < y \leq 0$.

$-2 \leq x < -1 \Rightarrow 1 < x^2 \leq 4 \Rightarrow -4 \leq -x^2 < -1 \Rightarrow -4 \leq y < -1$.

Thus, the range is: $-0.01 < y \leq 0$ or $-4 \leq y < -1$.

D. Find the range of a function $y = f(x) = 2x^2$ for x real.

That x is real means that any real number can be the input.
That is, the domain is a set of all real numbers. So f holds for all real numbers.

Whatever a real number it may be, the square of it is ≥ 0.
And the expression is $2x^2$, so each output is twice the square of each input.

Nevertheless, every output is still greater than or equal to 0.
Therefore, the range is: $y \geq 0$.

E. Find the range of a function $y = f(x) = 2x^2$ for $x \geq 0$.

The domain is: $x \geq 0$. So f holds for $x \geq 0$. And we have: $y = f(x) = 2x^2$.

So we get: $x \geq 0 \Rightarrow 2x^2 \geq 0 \Rightarrow y \geq 0$. Therefore, the range is: $y \geq 0$.

F. Find the range of a function $y = f(x) = 2x^2$ for $x \leq 0$.

The domain is: $x \leq 0$. So f holds for $x \leq 0$. And we have: $y = f(x) = 2x^2$.

So we get: $x \leq 0 \Rightarrow x^2 \geq 0 \Rightarrow 2x^2 \geq 0 \Rightarrow y \geq 0$. Therefore, the range is: $y \geq 0$.

G. Find the range of a function $y = f(x) = 2x^2$ for $0 \leq x \leq 1$.

The domain is: $0 \leq x \leq 1$. Consequently, f holds for $0 \leq x \leq 1$. And we have: $y = 2x^2$.

So we get: $0 \leq x \leq 1 \Rightarrow 0 \leq x^2 \leq 1 \Rightarrow 0 \leq 2x^2 \leq 2 \Rightarrow 0 \leq y \leq 2$, which is the range.

H. Find the range of a function $y = f(x) = 2x^2$ for $x \geq 1$.

The domain is: $x \leq 0$. So f holds for $x \leq 0$. And we have: $y = f(x) = 2x^2$.

So we get: $1 \leq x \Rightarrow 1 \leq x^2 \Rightarrow 2 \leq 2x^2 \Rightarrow y \geq 2$, which is the range.

I. Find the range of a function $y = f(x) = -2x^2$ for $x \geq -1$.

The domain is: $x \geq -1$. So f holds for $x \geq -1$. And we have: $y = f(x) = -2x^2$.

So we get: $-1 \leq x \Rightarrow 0 \leq x^2$ since x can be 0, and thus, we get:

$0 \leq 2x^2 \Rightarrow 0 \geq -2x^2 \Rightarrow y \leq 0$, which is the range.

J. Find the range of a function $y = f(x) = -2x^2$ for $x \leq 1$.

The domain is: $x \leq 1$. So f holds for $x \leq 1$. And we have: $y = f(x) = -2x^2$.

So we get: $1 \geq x \Rightarrow 0 \leq x^2$ since x can be 0, and thus, we get:

$0 \leq 2x^2 \Rightarrow 0 \geq -2x^2 \Rightarrow y \leq 0$, which is the range.

K. Find the range of a function $y = f(x) = -2x^2$ for $1 < x \leq 2$.

The domain is: $1 < x \leq 2$. So f holds for $1 < x \leq 2$. And we have: $y = f(x) = -2x^2$.

So we get: $1 < x \leq 2 \Rightarrow 1 < x^2 \leq 4 \Rightarrow -4 \leq -x^2 < -1 \Rightarrow -8 \leq -2x^2 < -2 \Rightarrow -8 \leq y < -2$.

Therefore, the range is: $-8 \leq y < -2$.

L.   Find the range of a function $y = f(x) = 2x^2 + x$ for $1 \le x \le 2$ or $-2 \le x \le -1$.

The domain is: $1 \le x \le 2$ or $-2 \le x \le -1$. So f holds for $1 \le x \le 2$ or $-2 \le x \le -1$.

And we have: $y = f(x) = 2x^2 + x$, which is not simple quadratic as x^2 or $3x^2$.
So it may not be simple enough to apply inequalities only to find the range.

That's because it's not quite simple to keep track of the output, because the output does not usually change constant manner. That is to say that it is often the case, it does not keep increasing or decreasing. So it's not a good idea to apply inequalities only to track the outputs changing as the inputs change. What then, is a good idea?

In such a case in fact, we can get the range easily and fast using the graph of f.
More importantly though, it is safe to get the range graphically.
So it's not only easy and fast but safe, too, to find the range using a graph.

So graphing matters.

Let's one more time though, find the range applying inequalities.
That is, we are going to get the range tacking the outputs changing as the inputs change.

To begin with, the domain is made of two intervals:

One is $1 \le x \le 2$, and the other is: $-2 \le x \le -1$. Thus, we have two cases.

First: $1 \le x \le 2 \Rightarrow 1 \le x^2 \le 4 \Rightarrow 2 \le 2x^2 \le 8$, so one case is: $1 \le x \le 2$ and $2 \le 2x^2 \le 8$.

Next: $-2 \le x \le -1 \Rightarrow 1 \le x^2 \le 4 \Rightarrow 2 \le 2x^2 \le 8$, so the other is: $-2 \le x \le -1$ and $2 \le 2x^2 \le 8$.

And we have: $y = 2x^2 + x$, which is the sum of $2x^2$ and x.

So beginning with the case where $1 \le x \le 2$ and $2 \le 2x^2 \le 8$, we can see that:

• First, when x is 1, $2x^2$ is 2, and when x is 2, $2x^2$ is 8.

• Next, if the value of x increases from 1 to 2, the value of $2x^2$ increases from 2 to 8.

So the value of $(2x^2 + x)$ keeps increasing as the value of x changes from 1 to 2.

Thus, the smallest value of $(2x^2 + x)$ is the sum of the minimums of $2x^2$ and x, and the greatest is the sum of the maximums of the two.

So next, let's get the maximums and minimums.

To begin with, we have: $1 \leq x \leq 2$, so the minimum of x is 1, and the maximum is 2.

And next, we have: $2 \leq 2x^2 \leq 8$, so the minimum of $2x^2$ is 2, and the maximum is 8.

Thus, we get: $2 + 1 \leq 2x^2 + x \leq 8 + 2 \Rightarrow 3 \leq 2x^2 + x \leq 10$.

So we get: $3 \leq 2x^2 + x \leq 10$ for $1 \leq x \leq 2$. Thus, we get: $3 \leq y \leq 10$ for $1 \leq x \leq 2$.

And we can put the idea above the way below, too:

We know that the smallest value of $(2x^2 + x)$ is the sum of the minimums of $2x^2$ and x, and that the biggest value of $(2x^2 + x)$ is the sum of the maximums of $2x^2$ and x.

And we have: $y = f(x) = 2x^2 + x$ for $1 \leq x \leq 2$.

So we can say that when x is 1, the value of $f(x)$ is the smallest, and that when x is 2, the value of $f(x)$ is the biggest.

That is to say that the smallest value of $f(x)$ is $f(1)$, and that the biggest value of $f(x)$ is $f(2)$. So we can put it this way: $f(1) \leq f(x) \leq f(2)$.

And we have: $f(1) = 2 \cdot 1^2 + 1 = 3$, and $f(2) = 2 \cdot 2^2 + 2 = 10$.

Thus, we get: $3 \leq f(x) \leq 10$. And we have: $y = f(x)$. So we get: $3 \leq y \leq 10$ for $1 \leq x \leq 2$.

Next, moving on to the second case where $-2 \leq x \leq -1$ and $2 \leq 2x^2 \leq 8$, we can see that:

• First, when x is -2, $2x^2$ is 8, and when x is -1, $2x^2$ is 2.

• Next, if the value of x increases from -2 to -1, the value of $2x^2$ decreases from 8 to 2.

And when one thing decreases, if we keep adding a negative to it, it keeps decreasing.

So the value of $(2x^2 + x)$ decreases constantly as x changes from -2 to -1.

And we have: $y = f(x) = 2x^2 + x$ for $-2 \leq x \leq -1$.

So we can say that when x is -2, the value of $f(x)$ is the greatest, and that when x is -1, the value of $f(x)$ is the smallest.

That is to say that the biggest value of $f(x)$ is $f(-2)$, and that the smallest value of $f(x)$ is $f(-1)$. So we can put it this way: $f(-2) \geq f(x) \geq f(-1)$.

And we have: $f(-1) = 2 \cdot (-1)^2 + (-1) = 1$, and $f(-2) = 2 \cdot (-2)^2 + (-2) = 6$.

Thus, we get: $1 \leq f(x) \leq 6$. And we have: $y = f(x)$. So we get: $1 \leq y \leq 6$ for $-2 \leq x \leq -1$.

Now, putting threads together, we have:

$3 \leq y \leq 10$ for $1 \leq x \leq 2$, and $1 \leq y \leq 6$ for $-2 \leq x \leq -1$.

That is, when $1 \leq x \leq 2$ or $-2 \leq x \leq -1$, we get: $3 \leq y \leq 10$ or $1 \leq y \leq 6$.

And notice that an interval where $3 \leq y \leq 6$ is common to $3 \leq y \leq 10$ and $1 \leq y \leq 6$.

So we get: $1 \leq y \leq 10$ for $1 \leq x \leq 2$ or $-2 \leq x \leq -1$.

Therefore, the range is: $1 \leq y \leq 10$.

As stated earlier though, finding ranges of functions nonlinear, we may not want to approach the ranges the way above. We may not want to just apply inequalities only.

Usually, using the graph of a function, you can get the range easily and fast. More importantly though, you can get it safely. You don't want to get the wrong solution readily and quickly, do you?

So graphing matters. And it does a lot.

(Graphing is covered in the series of **CONICS**. And the operations with graphs or curves are covered in **GRAPH OPERATIONS**.)

So let's now try getting the range considering the nature of the curve of f.

Examining the expression where $2x^2 + x$, we can see it's <u>quadratic</u>, and the coefficient of the quadratic term is 2, which is *positive*, so the curve is a <u>parabola</u>, and is *concave-up*.

So since the parabola is concave-up, as the value of x increases on the left of the *axis of symmetry*, the value of $f(x)$, that is, the value of y keeps decreasing, and as the value of x increases on the right of the axis of symmetry, the value of $f(x)$ keeps increasing.

(If not sure of such a coefficient and such an axis of symmetry, refer to **CONICS 2**.)

And we have: $y = f(x) = 2x^2 + x$ for $1 \leq x \leq 2$ or $-2 \leq x \leq -1$.

Of this parabola, the axis of symmetry is: $x = -\frac{1}{4}$.
The axis of symmetry of a parabola $y = ax^2 + bx + c$ is a line: $x = -\frac{b}{2a}$.

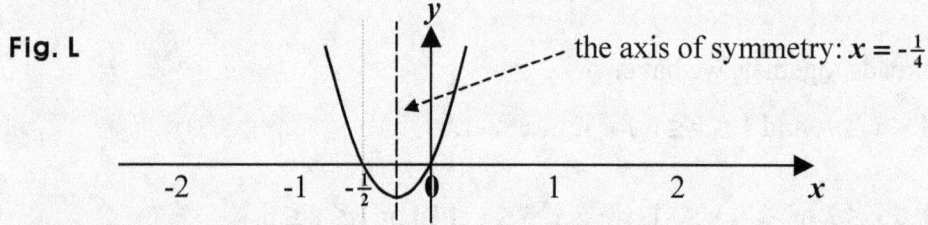

Fig. L   the axis of symmetry: $x = -\frac{1}{4}$.

So the interval where: $-2 \leq x \leq -1$ is on the left of the axis of symmetry, which is: $x = -\frac{1}{4}$.
Thus, we can simply get: $-2 \leq x \leq -1 \Rightarrow f(-2) \geq f(x) \geq f(-1)$.

And the other interval where: $1 \leq x \leq 2$ is on the right of the axis of symmetry.
Thus, we can simply get: $1 \leq x \leq 2 \Rightarrow f(1) \leq f(x) \leq f(2)$.

So next, we want to get the values of $f(1), f(2), f(-2)$, and $f(-1)$.

Then, we get: $f(1) = 2x^2 + x = 2 + 1 = 3$, and by the same token, we get:
$f(2) = 8 + 2 = 10$, $f(-2) = 8 - 2 = 6$, and $f(-1) = 2 - 1 = 1$.

So we get: $f(1) \leq f(x) \leq f(2) \Rightarrow 3 \leq f(x) \leq 10$, and $f(-1) \leq f(x) \leq f(-2) \Rightarrow 1 \leq f(x) \leq 6$.

And we have: $y = f(x)$.   So we get: $3 \leq y \leq 10$ or $1 \leq y \leq 6$.

And $3 \leq y \leq 10$ is common to $3 \leq y \leq 10$ and $1 \leq y \leq 6$. Therefore, the range is: $1 \leq y \leq 10$.

Practice 4 on Domains and Ranges

These are also, for familiarity with functions, particularly, domains and range, and also, for your algebra on inequalities, not quite straightforward sometimes.

Find the range of each function as follows.

0. $y = f(x) = -x^2 + 2x$ for $0 \le x \le 1$ or $-2 \le x \le -1$.

1. $y = f(x) = -x^2 + 2x$ for $0 \le x \le 3$.

2. $y = f(x) = -2x^2$ for $0 \le x < 0.1$ or $-2 \le x \le -1$.

3. $y = f(x) = \frac{1}{x^2}$ for x nonzero real.

4. $y = f(x) = -\frac{1}{x^2}$ for $x > 0$.

5. $y = f(x) = -\frac{1}{x^2}$ for $-1 \le x < 0$ or $x > 0$.

Suggestions or Solutions
To the Problems in Practice 4

0. Find the range of a function $y = f(x) = -x^2 + 2x$ for $0 \leq x \leq 1$ or $-2 \leq x \leq -1$.

As stated earlier in the practice set 3, when finding ranges of functions nonlinear, we may not want to apply inequalities only. And this is particularly the case.
So we don't want to apply inequalities only doing this example.

Usually, using the graph of a function, you can get the range readily and quickly.
And more importantly, you can get it safely. So graphing matters. And it does a lot.

If therefore, familiar with the nature of the curve of the function, we can get the range fast and easily, too, doing just simple algebra looking at the curve.

So let's now try getting the range considering the nature of the curve of f.

Examining the expression where $-x^2 + 2x$, we can see it's <u>quadratic</u>, and the coefficient of the quadratic term is -1, which is *negative*, so the curve is a <u>parabola</u>, *concave-down*.

So since the parabola is concave-down, as the value of x increases on the left of the *axis of symmetry*, the value of $f(x)$, that is, the value of y keeps increasing, and as the value of x increases on the right of the axis of symmetry, the value of $f(x)$ keeps decreasing.

(If not sure of such a coefficient and such an axis of symmetry, refer to **CONICS 2**.)

And we have: $y = f(x) = -x^2 + 2x$ for $0 \leq x \leq 1$ or $-2 \leq x \leq -1$.

Of this parabola, the axis of symmetry is: $x = 1$.

The axis of symmetry of a parabola $y = ax^2 + bx + c$ is a line $x = -\frac{b}{2a}$.

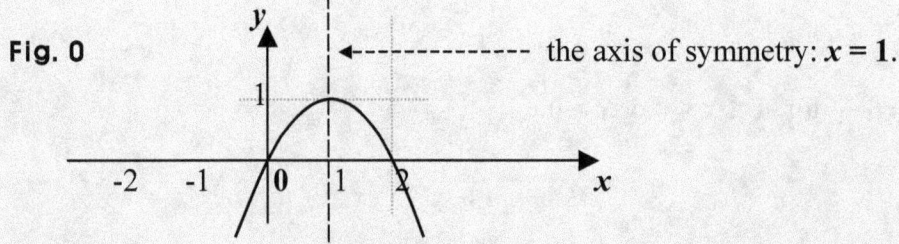

Fig. 0   the axis of symmetry: $x = 1$.

So the interval where: **-2 ≤ x ≤ -1** is on the left of the axis of symmetry, which is: $x = 1$.

Thus, we can simply get: **-2 ≤ x ≤ -1 ⇒ f(-2) ≤ f(x) ≤ f(-1)**.

And the other interval where: **0 ≤ x ≤ 1** is still on the left of the axis of symmetry.

Thus, we can simply get: **0 ≤ x ≤ 1 ⇒ f(0) ≤ f(x) ≤ f(1)**.

So next, we want to get: **f(0), f(1), f(-2), and f(-1)**.

Then, we get: **f(0) = -x² + 2x = 0 + 0 = 0**, and by the same token, we get:

f(1) = -1 + 2 = 1,   f(-2) = -4 - 4 = -8, and f(-1) = -1 - 2 = -3.

So we get: **f(-2) ≤ f(x) ≤ f(-1) ⇒ -8 ≤ f(x) ≤ -3, and f(0) ≤ f(x) ≤ f(1) ⇒ 0 ≤ f(x) ≤ 1**.

And we have: **y = f(x).   So we get: -8 ≤ y ≤ -3 or 0 ≤ y ≤ 1**.

Therefore, the range is: **-8 ≤ y ≤ -3 or 0 ≤ y ≤ 1**.

1.   **Find the range of a function y = f(x) = -x² + 2x for 0 ≤ x ≤ 3.**

As stated earlier, when finding ranges of functions nonlinear, we may not want to apply inequalities only. And this is particularly the case, and is in fact, worse than the one in the example 0 above. So we don't want to apply inequalities only, doing this example.

And as mentioned earlier, too, if familiar with the nature of the curve and the expression of the function, we can get the range fast and easily, too, doing just simple algebra, actually looking at the curve.

So let's now try getting the range considering the nature of the curve of *f*.

Examining the expression where **-x² + 2x**, we can see it's <u>quadratic,</u> and the coefficient of the quadratic term is -1, which is *negative*, so the curve is a <u>parabola,</u> *concave-down*.

So since the parabola is concave-down, as the value of *x* increases on the left of the *axis of symmetry*, the value of *f(x)* keeps increasing, and as the value of *x* increases on the right of the axis of symmetry, the value of *f(x)* keeps decreasing.

And we have: $y = f(x) = -x^2 + 2x$ for $0 \leq x \leq 3$, where the axis of symmetry is:

$x = 1$, because the axis of symmetry of a parabola $y = ax^2 + bx + c$ is a line $x = -\frac{b}{2a}$.

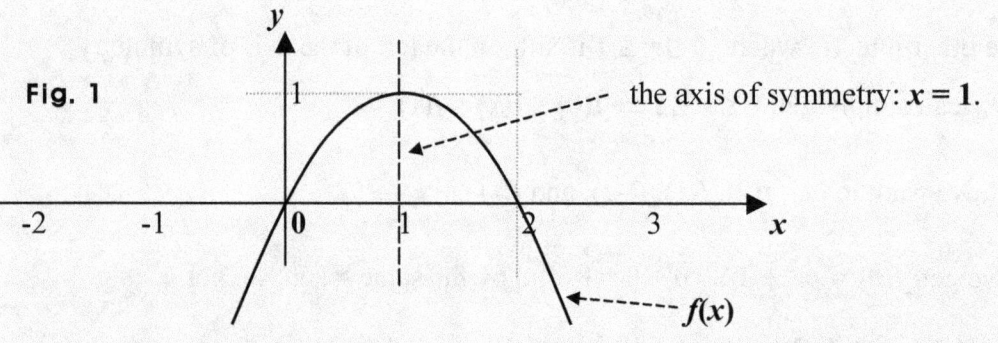

Fig. 1   the axis of symmetry: $x = 1$.

So in this case, the interval: $0 \leq x \leq 3$ is not on either side, that is, the axis of symmetry, that is the line $x = 1$ is passing through between both ends of the interval.   So?

We know that the parabola is concave-down. So as x changes from 0 to 3, $f(x)$ increases up to the point called the vertex, and then, decreases after the vertex. The vertex is thus, the highest point in the parabola.

So we want to get three values, one is $f(0)$, and another is $f(3)$.   What then, is the other?

It is the y-coordinate at the vertex. So it is $f(1)$, because the axis of symmetry passes through the vertex, and the axis is a line $x = 1$, so the vertex is at $(1, f(1))$.

Since the parabola is concave-down, the value of $f(1)$ is the greatest value of the function $f(x)$. So it can let us know one end of the interval that is the range of the function f. Finding thus, the three values, we can get first:

$f(0) = -x^2 + 2x = 0 + 0 = 0$, and by the same token, we get: $f(1) = -1 + 2 = 1$, and $f(3) = -9 + 6 = -3$.

So for $0 \leq x \leq 3$, we get: $f(3) \leq f(x) \leq f(1) \Rightarrow -3 \leq f(x) \leq 1$

And we have: $y = f(x)$.   So we get: $-3 \leq y \leq 1$, which is the range.

2. **Find the range of a function $y = f(x) = -2x^2$ for $0 \le x < 0.1$ or $-2 \le x \le -1$.**

Though the expression, that is, the function is not linear in this case, it is quite simple. Anyway, the axis of symmetry is the y-axis, that is, a line $x = 0$. So the axis of symmetry is just at one end of the domain, not within the domain, and thus, should not cause much trouble in finding the range. So let's try this time, getting the range applying inequalities.

The domain is: $0 \le x < 0.1$ or $-2 \le x \le -1$.   So f holds for $0 \le x < 0.1$ or $-2 \le x \le -1$.
And we have: $y = f(x) = -2x^2$.   So we get:

$$0 \le x < 0.1 \Rightarrow 0 \le x^2 < 0.01 \Rightarrow -0.01 < -x^2 \le 0 \Rightarrow -0.02 < -2x^2 \le 0 \Rightarrow -0.02 < y \le 0.$$

$$-2 \le x \le -1 \Rightarrow 1 < x^2 \le 4 \Rightarrow -4 \le -x^2 < -1 \Rightarrow -8 \le -2x^2 < -2 \Rightarrow -8 < y \le -2.$$

Thus, we get: $-0.02 < y \le 0$ for $0 \le x < 0.1$, and also, $-8 \le y < -2$ for $-2 \le x \le -1$.
Therefore, the range is: $-0.02 < y \le 0$ or $-8 \le y < -2$.

3. **Find the range of a function $y = f(x) = \frac{1}{x^2}$ for x nonzero real.**

That x is nonzero real means that the input can be any real number other than 0.
In fact, the function cannot hold for 0 since the denominator is 0 if $x = 0$.

So the domain can be put this way: $x < 0$ or $x > 0$.
Therefore, the function f holds for $x < 0$ or $x > 0$.

Considering the nature of the expression however, we may want to put the domain this way: $0 < |x| < 1$, and $|x| \ge 1$.

Then first, if the magnitude of x is between 0 and 1, x^2 is between 0 and 1, too.
That is, we have: $0 < |x| < 1 \Rightarrow 0 < x^2 < 1$.   Thus, we get: $\frac{1}{x^2} > 1$. How come?

If for instance, $x^2 = \frac{1}{4}$, we get: $\frac{1}{x^2} = 4$.   Thus, we get: $0 < |x| < 1 \Rightarrow \frac{1}{x^2} > 1$.
And we have: $y = f(x) = \frac{1}{x^2}$. So the range is: $y > 1$ if the domain is: $0 < |x| < 1$.

And next, if $|x| \geq 1$, we get: $x^2 \geq 1$.

So if $|x| \geq 1$, we get: $0 < \frac{1}{x^2} \leq 1$.   And we have: $y = f(x) = \frac{1}{x^2}$, too.

So we get: $0 < y \leq 1$ for $|x| \geq 1$.   And we have: $y > 1$ for $0 < |x| < 1$, too.

Putting thus, threads together, we have: $y > 1$ or $0 < y \leq 1$. So the range is: $y > 0$.

4.   Find the range of a function $y = f(x) = -\frac{1}{x^2}$ for $x > 0$.

The domain is: $x > 0$. So the function f is defined for $x > 0$.

Though the domain is just one interval, we want to consider two cases.   Why?

It's because of the nature of the expression. And the two cases are: $0 < x < 1$, and $x \geq 1$.

Now first, if x is between 0 and 1, x^2 is between 0 and 1, too.

That is, we have: $0 < x < 1 \Rightarrow 0 < x^2 < 1$.   So we get: $\frac{1}{x^2} > 1$ when $0 < x < 1$.

We have: $y = f(x) = -\frac{1}{x^2}$, though.   Thus, we get: $y = -\frac{1}{x^2} < -1$ for $0 < x < 1$.

So the range is: $y < -1$ if the domain is: $0 < x < 1$.

And next, if $x \geq 1$, we get: $x^2 \geq 1$.   So we get: $0 < \frac{1}{x^2} \leq 1$, since $\frac{1}{x^2}$ cannot be ≤ 0.

So if $x \geq 1$, we get: $0 < \frac{1}{x^2} \leq 1$.   We have: $y = f(x) = -\frac{1}{x^2}$, though.

Thus, we get: $-1 \leq -\frac{1}{x^2} < 0$.

So we get: $-1 \leq y < 0$ for $x \geq 1$.   And yet, we have: $y < -1$ for $0 < x < 1$, too.

Thus, we have: $-1 \leq y < 0$ or $y < -1$.   So we get: $y < 0$, and therefore, the range is: $y < 0$.

And in fact, putting in a graph the curve of $y = f(x) = \frac{1}{x^2}$ for x nonzero real, we get:

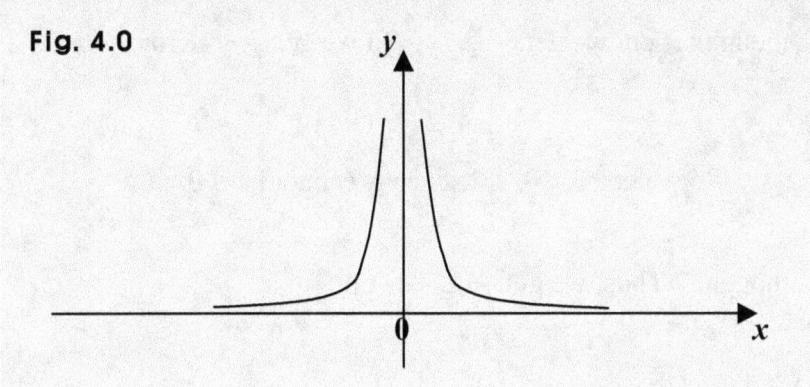

Fig. 4.0

And putting in a graph the curve of $y = f(x) = -\frac{1}{x^2}$ for $x > 0$, we get:

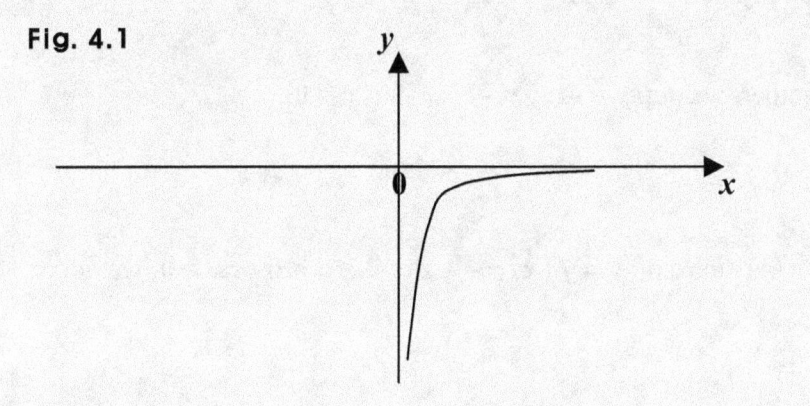

Fig. 4.1

5.   Find the range of a function where $y = f(x) = -\frac{1}{x^2}$ for $-1 \leq x < 0$ or $x > 0$.

The domain is: **$-1 \leq x < 0$ or $x > 0$.**   So f is defined for **$-1 \leq x < 0$ or $x > 0$.**

Though the domain is made of two intervals, yet we may want to consider three cases.

That is because of the nature of the expression.   And the three cases are as follows:

• **$-1 < x < 0$ or $0 < x < 1$**, that is, the magnitude of x is between 0 and 1, so **$0 < |x| < 1$.**

• **$x \geq 1$.**

• **$x = -1$.**

Then first, if $0 < |x| < 1$, we get: $0 < x^2 < 1$.   So we get: $\frac{1}{x^2} > 1$.

We have: $y = f(x) = -\frac{1}{x^2}$, though. Then, we get: $-\frac{1}{x^2} < -1$. So we get: $y < -1$ for $0 < |x| < 1$.

Next, if $x \geq 1$, we get: $x^2 \geq 1$.   So we get: $0 < \frac{1}{x^2} \leq 1$, since $\frac{1}{x^2}$ cannot be ≤ 0.

We have: $y = f(x) = -\frac{1}{x^2}$, though.   Thus, we get: $-1 \leq -\frac{1}{x^2} < 0$.

So we get: $-1 \leq y < 0$ for $x \geq 1$.

And next, if $x = -1$, we simply get: $-\frac{1}{x^2} = -1$.

Now, putting threads together, we get: $y = -1$, $y < -1$, or $-1 \leq y < 0$.

Thus, the range is: $y < 0$.

And, putting in a graph the curve of $y = f(x) = -\frac{1}{x^2}$ for $-1 \leq x < 0$ or $x > 0$, we get:

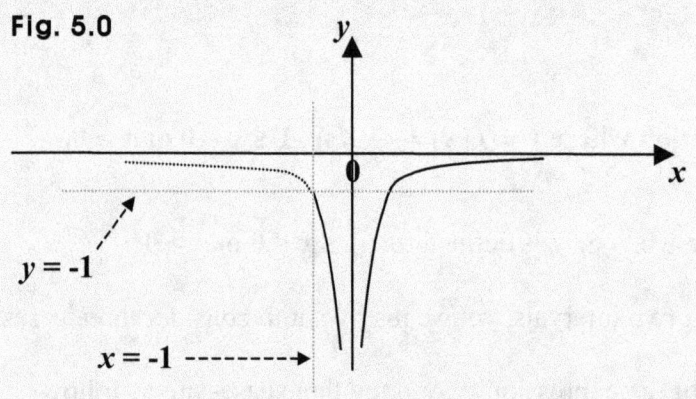

Fig. 5.0

$y = -1$

$x = -1$ -------->

1. Why variables?

Working with functions, we usually work with variables, because functions are math expressions, which are usually made of variables, together with numbers, of course. How come though, they are said to be *variable*, and what's the use?

Doing math, we work with values.   What values, though?

All kinds of values. Doing math, we do operations with values. Adding for instance, we do additions. Doing such operations, we get values, too.   What values though?

Quite ambiguous, isn't it?

So we don't just do such operations. We use tools called *numbers* to specify values. So adding for instance, we add numbers together, and get the sum.   So?

A number has a value.   How does it get a value, though?

We assign a value to it.
So for instance, we assign seven to 7, which has thus, a value called seven. And that's it.

For another instance, 1 has a value called one, and is thus, of course, read as one.

And adding 7 and 1 together, we get the sum. Then, the sum gets a value called eight.

We have 8 that has a value called eight. So we can say that the sum is 8.

We have 2, also, so adding 2 and 7 together, we get the sum, which is 9. So this time, the sum is 9. Until we do another addition, the sum remains 9.

And we can give a name to the sum. So we can name it using a letter, and it can be called *s*, for instance. Then, *s* can change its value every time the sum gets a new value.

Every time therefore, we add two numbers together, the sum gets a new value, so *s* gets the new value. Until however, we get the next sum, *s* has to keep the new value. That is to say that the value of *s* has to remain the same until it is assigned another new value.

In other words, the value of *s* has to be constant until we get the next sum.
In short, we say that *s* is *constant*.

And *s* is an object, so we just call *s* *a constant*.
So a constant is an object that keeps a value until it gets assigned a new value.

And we can make as many constants as we need.   How then, do we make them?

We just assume those, or declare some letters constants, in proper manner, of course.

So for instance, we can set up an operation this way: $a + b = c$ where *a*, *b*, and *c* are constants. Or just setting: $a + b = c$, we normally assume that *a*, *b*, and *c* are constants.

Then, giving a value to each of *a* and *b*, we get the sum, which is the value of *c*.

So we have tools called *numbers* as 0, 1, 2, 3, etc., and tools called *constants* as *a*, *b*, etc. Doing math therefore, we can conveniently use those tools doing operations as additions. What is the difference though, between a number and a constant?

A number cannot change its value once it's been given a value.

So for instance, a number 7 has a value called seven forever, and thus, we cannot change the value of 7. That's because we do not give a new value to a number with a value.

A constant can however, change its value if we give a new value to it.
So we can give a new value to a constant with a value.
It has to keep though, a certain value until it gets assigned a new value.

So we can readily show a math operation in a general manner using such tools called constants. For instance, setting: $\underline{a = 2b}$ where $\underline{a}$ and $\underline{b}$ are constants, we can say that the value of a is twice the value of b no matter what the value of b may be if $b \neq 0$.

How then, do we know if c is a constant of a *particular kind*, for instance?

We can define it to be such a constant by declaring it a constant of the particular kind.

So for instance, declaring c a constant *integer*, we take c for an integer, and can use it as if it were an integer. And *at a time*, we can assign one number (value) only to a constant. So we can assign one integer to c, and can use c as the integer.

Thus, for instance, c can get one of -2, 0, 1, 3, etc.
It has to however, keep the integer given until we assign another integer to it, so it cannot change its value by itself. So for instance:

Assuming now, c is given 3, we use c as 3.
And if we want to give -1 to c, we give -1 to c, and use c as -1, and of course, we do not use c as 3.
If however, we want to give 3 to c again, we give 3 to c, and use c as 3, and of course, we do not use c as -1.

For another instance, declaring d a constant *real*, we can just declare d a constant.
So just declaring d a constant, we can use d as a real number, and can assign at a time, one real number to d, which has to keep the real number assigned until we assign another real number to it.

Suppose for instance, we get an equation: $\underline{y = dx^2 + 1}$ where $\underline{d}$ is a nonzero constant.

Then, we use *d* as a nonzero real number, and *d* can be given one real number nonzero, but has to keep the real number given until it gets assigned another real number nonzero.

So the equation above can be said to represent all the equations as $y = x^2 + 1$, $y = 2x^2 + 1$, $y = -x^2 + 1$, $y = x^2/2 + 1$, etc.

Suppose for another instance, we get an equation: $\underline{y = bx^2}$ where *b* is a constant integer.

Then, *b* is an integer constant. So we use *b* as an integer, and *b* can get one integer, but has to keep the integer given until we give another integer to it.

Suppose for another instance, we get a statement below:

• Assuming that *a* is constant, find the point where a line $y = 2x + a$ meets the *x*-axis.

Then, we use *a* as a real number, and *a* can represent all real numbers, but has to keep a real number assigned until it gets assigned another real number.

Let's take one more example. Suppose we get a statement as below:

Assuming *d* is a constant > 3, and *A* is a line $y = x + d$, find the point where *A* meets the *x*-axis.

Then, we use *d* as a real number > 3, so *d* can be assigned one real number > 3, but has to keep the real number assigned until it gets assigned another real number > 3.

So a constant itself cannot change its value, but we can change the value of a constant if we need to.

And we have another kind in tools that work like numbers or constants, and thus can have values.

Such a tool is very different from a number, but seems quite close to a constant. It's certainly not a constant, of course.   What then, is the tool?

It *looks* no other than a constant.

So like a constant, it can have one value at a time, and cannot have two or more values at a time. Unlike a constant though, it keeps changing its value. Or rather, it has to constantly change its value.

So its value varies continuously. And thus, we call it *a variable*.

So a variable varies its value continuously, and thus, its value keeps changing. What values then, can a variable have?

A variable covers all values in a set. For instance, it can cover a set of all values ≥ 0.

And we use numbers to show values. So for instance, a variable can cover a set of all real numbers, a set of all nonzero real numbers, or a set of all real numbers less than 1.

And for instance, if a variable covers all real numbers positive, we only know the fact that the variable keeps changing its value covering all positive real numbers.

However, a variable cannot have two or more numbers at a time, and has to have one number only at each moment.

How then, can we get a variable? That is, how can we make one?

Like other tools in math, defining a particular variable, we can make one. So by a variable definition, we can make one and use it.

A variable varies covering all the values in a particular kind or set, which is a number set, of course. So we can define one by declaring it that way. So let's now make one.

Note:

We can use '$\in$' to show that particular objects *belong to* or *are elements of* other object. For instance, setting: $a \in B$, we mean that a belongs to or is an element of B, and also, setting: u and $v \in W$, we mean that u and v both belong to or are elements of W.

Suppose now, for instance, we get a statement as follows: **_d_ is real**.

Then, the statement is a variable definition, and it is assumed that **_d_** is a variable that can take any real number, and covers all real numbers.

So **_d_** keeps changing its value, and has one real number at each moment.

Suppose for another instance, we get an *equation* as follows: $y = x^2 + x + 1$.

Then, it is assumed that **_x_** is a variable that can take any real number, and covers all real numbers, because **_x_** can get any real number in this case. What then, about **_y_**?

It is a variable, too, and covers a set of some real numbers. And if no declaration is done to the letters as **_x_** and **_y_**, we just assume that the letters are variables.

So **_x_** and **_y_** keep changing their values, and each has one real number at a time.

And we can notice that the values of the variables **_x_** and **_y_** are dependent upon each other. That is, the value of one variable determines the value of the other.

So for instance, when **_x_** is 1, that is, if $x = 1$, we get: $y = 1 + 1 + 1 = 3$.
And also, when **_x_** is -2, that is, if $x = -2$, we get: $y = (-2)^2 + (-2) + 1 = 3$, too.
In other words, if $y = 3$, we get: $x = 1$ or -2.

And in fact, the value of **_y_** is limited in this case. That's because we get:

$$y = x^2 + x + 1 = x^2 + x + \tfrac{1}{4} - \tfrac{1}{4} + 1 = (x + \tfrac{1}{2})^2 + \tfrac{3}{4} \geq \tfrac{3}{4}. \quad \text{So we get: } y \geq \tfrac{3}{4}.$$

And thus, we can say that **_y_** is a variable that covers all the numbers $\geq \tfrac{3}{4}$.

Also, we can limit the value a variable can take this way, too: $y = x^2 + x + 1$ for $x > 1$.

Then, it is assumed that **_x_** is a variable that can take any real number > 1.
What then, about **_y_**?

It is assumed that y is a variable, too, of course, but all the real numbers y can take are subject to the condition that $x > 1$. So let's now find the set of all the numbers y can take.

To begin with, we have: $y = x^2 + x + 1 = (x + \frac{1}{2})^2 + \frac{3}{4}$. And we have: $x > 1$.

So we get: $y > (1 + \frac{1}{2})^2 + \frac{3}{4} = \frac{9}{4} + \frac{3}{4} = 3$. Thus, we get: $y > 3$.

So it is assumed that y is a variable that can take any real number > 3.

If defining a constant though, we need to declare it a constant. So for instance, if we just get: $y = cx^2 + x + 1$, we have to assume that not only x and y but c is a variable, too.

If however, we get: $\underline{y = cx^2 + x + 1 \text{ where } c \text{ is a constant}}$, we just assume that x and y only are variables.

And for another instance, declaring: $\underline{0 < c < 1 \text{ where } c \text{ is constant}}$, we mean that c is a constant that can be assigned at a time, one real number between 0 and 1.

Note that we can change the value of a constant if we want to, so a constant itself cannot change its value unlike a variable, and thus, a number as 7 *can be* technically *taken for* a constant, too, since 7 cannot change its value anyway. Normally however, we don't call a number a constant, of course.

 • Let's next, define variables using set notation.

Setting: $\underline{A = \{e| \ 1 < e < 2\}, \text{ and } c \in A}$, we mean that c is a variable that can take any real number between 1 and 2. So it is just as good as setting: $1 < c < 2$.

For another instance, setting: $A = \{e| \ 1 < e < 2\}$, $B = \{t| \ 3 < t < 4\}$, and $s \in A$ or B, we mean s is a variable that can take any number between 1 and 2 or between 3 and 4. And of course, s cannot take any number ≤ 1, any number from 2 to 3, and any number ≥ 4.

So for instance, s can take 1.5 or 3.8, but cannot take 2.

Also, it is just as good as setting: $\underline{1 < s < 2 \text{ or } 3 < s < 4}$.

- Let's this time, define constants using set notation.

Stating: $A = \{e \mid 1 < e < 2\}$, and c is a constant $\in A$, we mean that c is a constant that can take at a time, one real number between 1 and 2.

So it is just as good as declaring or stating: $1 < c < 2$ where c is constant.

And next, stating: $A = \{e \mid 1 < e < 2\}$, $B = \{t \mid 3 < t < 4\}$, and v is a constant $\in A$ or B, we mean that v is a constant that can take at a time, one real number between 1 and 2 or between 3 and 4. And of course, v cannot take any number ≤ 1, any number from 2 to 3, and any number ≥ 4. So for instance, v can take 1.2 or 3.1, but cannot take 4 or 5.

Also, it is just as good as declaring: $1 < v < 2$ or $3 < v < 4$ where v is constant.

Now, why do we frequently have to make and use, or work with variables?

Doing math, we often work with objects that change. What change, though?

Such an object changes by changing the value of itself.
Suppose for instance, an apple is falling.
Then, the position of the apple changes.

And we can put it the way below, too:

The distance from the position where the apple starts falling changes.
In short, the distance changes.
Then, what do we need to keep track of the distance?

We need an object that can change its value.

More specifically, we need an object that can indicate a distance, and can keep changing its own value.   How then, can we get the object?

We may want to begin with naming the object, and for instance, can name it d.
Then, d is the distance, and keeps changing its value.   How then, do we call d?

We call d a variable, and can have d get all real numbers it can get.

Since d is a distance, d has to be greater than or equal to 0. So we want d to be able to get any real number ≥ 0. And thus, we can simply declare that $d \geq 0$.
How then, does d change?

The distance changes as time changes, so we need to keep track of the time, too. We don't just take the measurement of time though. We need an object that can indicate time, and can keep changing its own value.   So what do we need?

We need another variable that indicates time, and for instance, can name it t.
Since t is time, t has to be greater than or equal to 0. So we want t to be able to get any real number ≥ 0. And thus, we can simply declare that $t \geq 0$, too.

So we now can use d as the distance, which keeps changing its value along the flow of time called t, which keeps changing, of course, its value, too.

So we now have two variables, and one is subject to the other.

Suppose now, we want to see where the apple is at a particular moment of time.
That is, we want to know a particular value of d after a particular amount of time t.

What then, do we need to find such a value of d?

62

We need a relation particular between the distance and time.   That is, we want a specific correlation between *d* and *t*.   In other words, we need the *connection* between the two.

How then, can we get the connection?

We can find it doing some experiments and collecting data.   That is, taking distances after some amounts of time, we can come up with the connection.
How then, can we come up with it?

We know that the distance *d* keeps changing its value as the time *t* changes its value.

Finding thus, the math expression that is expressed in terms of *t*, and produces a value of *d* for each value of *t*, we get the connection.

For instance, the expression can be $4.9t^2$ or $t^2 + 3t + 1$.   What then, is the connection?

The connection is of course, *the expression* in terms of *t*.

We know that *the expression* produces a value of *d* for each value of *t*.

So setting: *d* = *the expression*, we get an equation, and we call in fact, the equation the *connective expression* between *d* and *t*.

And we call such an *expression* a function.

And we say that the distance *d* is a function of time *t*.

That's because the value of *d*, that is, the value of the function changes as time *t* changes.

And in fact, the function is: $\frac{1}{2}gt^2$, where *g* is a constant, and more specifically, is called the gravitational constant on earth, and is approximately 9.8.

So for each value of *t*, the value of $\frac{1}{2}gt^2$ is the value of the distance *d*.

And the connective expression between *d* and *t* is: $d = \frac{1}{2}gt^2$.

So we can now, define a function that specifies the correlation between the amount of distance *d* and the amount of time *t*.

And the inputs are the values of *t*, and the outputs are those of *d*.
What then, is the domain?

The domain of the function is the set of all the values of *t*, of course.
How then, can we get the domain?

Assuming that the apple is initially *h* meters away from the ground, we can say that the maximum distance the apple can travel is *h* meters. So finding *t* when *d* = *h*, we get:

$$d = \tfrac{1}{2}gt^2 \Rightarrow h = \tfrac{1}{2}gt^2 \Rightarrow t^2 = \tfrac{2h}{g} \Rightarrow t = \sqrt{\tfrac{2h}{g}}, \text{ since we have: } t \geq 0.$$

So the domain is: $0 \leq t \leq \sqrt{\tfrac{2h}{g}}$.

And thus, assuming the function is called *D*, we can set: $d = D(t) = \tfrac{1}{2}gt^2$ for $0 \leq t \leq \sqrt{\tfrac{2h}{g}}$.

And we know that the distance *d* can have values from 0 to *h*, and is the output variable. So the range is: $0 \leq d \leq h$.

And we can apply such an idea as above to somewhere else.

For instance, we can use a variable as a speed, inventory, amount of time, price, height, depth, volume, force, area, etc. In other words, we can take as a variable any object that has to change its value covering all the values defined. And we can keep track of its change. So a variable is a pretty tool for many uses, isn't it? We do math therefore, to make life easier.

And we want to note that a change does not just happen.

Where there is a change, there is a rate.   What change though?

It can be a change in distance, a change in area, etc.   What rate then?

It's a rate of change.   What rate of change though?

It usually is a time rate of change, and for instance, we often use a time rate of change in distance, which is called a velocity, which has a magnitude called a speed.

For instance, light travels at approximately three hundred thousand kilometers a second, which is a magnitude of a rate of change, and the magnitude is the speed.
So the speed of light is about 300,000 km/sec.

A rate of change can be however, not only a velocity but something else, too.

It can be an acceleration, which is a rate of change in a rate of change, if you will.
And it can be called in fact, a time rate of change in velocity.
So a velocity can change as time changes, and we call such a rate an acceleration.
And of course, an acceleration can change, too. There can be thus, a rate of change in acceleration, also. So an acceleration can be a function of time, too.

And we will get to see and work with a lot of those in the math called calculus.
Basically thus, calculus is about functions. Why then, calculus?

It's because we want to know how things change.

₂.Formal Definitions

Defining an object in math, we produce a description of it. And the description has specifics on it, and is based on rules, principles, or theories relevant.

And we usually use a formal definition. What is a definition formal though?

Even if a definition is not formal, it is not said to be informal.
What then, is a formal definition?

Defining an object in math, we mainly use symbols, and use them as many as possible.

So defining a function, too, we use symbols as many as possible. Thus, we minimize the use of natural language as English, and maximize the use of symbols.

And producing a definition the way above, we can say that we make a formal definition.

And also, we can call a *formal* definition a *symbolic* definition, too, because of heavy use of symbols.

And the more formal it is, the easier it gets.

The worst thing in learning things math is ambiguity, and we can reduce ambiguity by means of a formal definition. That is because it is concise and reduces the weaknesses in meaning or limitation of natural language, because we use symbols. So it can make a definition laconic, and remove room for ambiguity. We can therefore, alleviate the difficulty of understanding the definition.

And before we get into the discussion of formal definitions, we want to note and be familiar with the ideas below:

- We use '≡' to show that particular objects are identical to each other.
So for instance, $A \equiv B$ means that A and B are identical to each other.

- A symbol $\varnothing$ indicates an empty set, which has no element in itself and can be put in {}.

- And we use '⊆' to indicate that a particular set is a subset of another set.
For instance, $S \subseteq T$ says that a set S is a subset of a set T.

Let's now make some function definitions, that is, define or make some functions.

We can make functions making formal definitions of those functions.

To begin with, we can define a function called G the way as follows.

Suppose that:

The function designator is G.

$X = \{s| \ 1 \leq s \leq 2\}$, which is the domain, and $Y = \{t| \ 14 \leq t \leq 17\}$, which is the range.

And the expression is: $3x + 11$.

Then, $y = G(x)$ for $x \in X$.

The set of statements in the box above can be a function definition, and there is nothing wrong with the definition, the definition of the function G. It is quite verbose, though.

So making or defining a definition, we may want to use as many symbols as possible, and make use of words minimal. Then, we can make the definition clear and succinct.

So let's now try defining the function G with less use of words.

That is, we are making a formal (symbolic) definition of the function G.

Then, we can produce the formal definition of G either of the ways as follows:

$X = \{s|\ 1 \leq s \leq 2\}$, and $Y = \{t|\ 14 \leq t \leq 17\}$.
$G\colon X \longrightarrow Y$.
$y = G(x) = 3x + 11$ for $x \in X$.

$X = \{s|\ 1 \leq s \leq 2\}$, and $Y = \{3s + 11|\ s \in X\}$.
$G\colon X \longrightarrow Y$, and $y = G(x)$.

$X = \{s|\ 1 \leq s \leq 2\}$, and $Y = \{3s + 11|\ 1 \leq s \leq 2\}$.
$G\colon X \longrightarrow Y$, and $y = G(x)$.

So all the three definitions above are the same, and thus, indicate one same function as follows: $y = G(x) = 3x + 11$ for $1 \leq x \leq 2$.

(The definition above is in fact, another formal definition of G, and is the simplest.)

Let's anyway now, take a look at what each component of the definition is about.

To begin with, the domain is: $X = \{s|\ 1 \leq s \leq 2\}$, which is telling us X is a set of all the real numbers that are greater than or equal to 1 and less than or equal to 2.

Next, the range is: $Y = \{t|\ 14 \leq t \leq 17\}$, which is saying that Y is made of all the real numbers ≥ 14 and ≤ 17. So using symbols, we can make communication simple and fast. And describing a function, we can also say that a function is from a domain to a range.

So next, the expression below is saying that G is a function from X to Y.

$G\colon X \longrightarrow Y$. So we can readily and quickly see that the domain is X and the range is Y.

The arrow says that the values in X cause the values in Y.

That is to say that the function G produces the values in Y using the values of X.

So in a symbolic definition, we don't have to declare the domain and the range. That is, we can still distinguish each by means of the direction of an arrow.

So the statement, "$G: X \longrightarrow Y$." is clearly saying that G is a function from X to Y.

And next, the expression $y = G(x) = 3x + 11$ is saying that y is the output variable, gets each output, which belongs to Y, x is the input variable, gets each input from X, and the expression is: $3x + 11$, together with the name of the function, which is $G(x)$ or just G.

What do we mean by though, $\{3s + 11 | s \in X\}$ and $\{3s + 11 | 1 \leq s \leq 2\}$?

First of all, both are the same sets, and each is saying that it is a set of all the values that can be made from the expression $3s + 11$ for $1 \leq s \leq 2$.

So it is a set of all the values that are made from the expression $3x + 11$ for $1 \leq x \leq 2$, and thus, the set is the range, and is: $Y = \{e | 14 \leq e \leq 17\}$.

So $\{3s + 11 | s \in X\}$, $\{3s + 11 | 1 \leq s \leq 2\}$, and $\{e | 14 \leq e \leq 17\}$ are identical to each other.

That is, we have: $\{e | 14 \leq e \leq 17\} \equiv \{3s + 11 | 1 \leq s \leq 2\} \equiv \{3s + 11 | s \in X\}$, because we have $X = \{e | 1 \leq e \leq 2\}$.

Also, we can see from the fact above, $3s + 11$ is in fact, $3x + 11$, so we can say that we can choose any letter for a variable insofar as the consistency is maintained.

Now, from the definition of the function G, we can say that:

The function is called G, which is a function of x, and from the domain X, each input gets into x in the expression $3x + 11$, in which the operations get performed, then the output for the input gets produced, is put into y, and thus, is an element of the range Y.

Besides, we can put the definition of G in any of the ways below, too:

$X = \{u|\ 1 \le u \le 2\}$, $Y = \{v|\ 14 \le v \le 17\}$, $X \xrightarrow{\ G\ } Y$, and $y = G(x) = 3x + 11$.

$X = \{s|\ 1 \le s \le 2\}$, $Y = \{3t + 11|\ t \in X\}$, $X \xrightarrow{\ G\ } Y$, and $y = G(x)$.

$X = \{p|\ 1 \le p \le 2\}$, $Y = \{3q + 11|\ 1 \le q \le 2\}$, $X \xrightarrow{\ G\ } Y$, and $y = G(x)$.

In the definitions above, we put the function designator above the arrow, and other than that, they are all the same as the previous ones.

That is, $G: X \longrightarrow Y$ is the same as $X \xrightarrow{\ G\ } Y$.

Now, let's have a look at another example.

X and Y are sets of all real numbers, $F: X \longrightarrow Y$, and $y = F(x) = 2x + 1$.

R is a set of all real numbers, X and $Y \subseteq R$, $F: X \longrightarrow Y$, and $y = F(x) = 2x + 1$.

X is a set of real numbers, $Y = \{2x + 1|\ x \in X\}$, $F: X \longrightarrow Y$, and $y = F(x)$.

R is a set of all real numbers, $X \subseteq R$, $Y = \{2s + 1|\ s \in X\}$, $F: X \longrightarrow Y$, and $y = F(x)$.

R is a set of all real numbers, $X \subseteq R$, $Y = \{F(x)|\ x \in X\}$, $F(x) = 2x + 1$, and $y = F(x)$.

Though looking different from each other, all the five definitions above are the same. In each of the definitions above, we can see that:

F is a function from X to Y, so X is the domain of F, and Y is the range.
The input variable is x, and the output variable is y, and the expression is: $2x + 1$.

• Also, we often say that y is the *image* of x by the function $y = F(x)$.

So Y can be said to be the set of the images of all the elements in X by the function F.

And of course, the domain and range cannot be empty sets, that is, $X \neq \varnothing$, and $Y \neq \varnothing$.

Besides, we have another way where we can define a function.

We can express directly the expression, along with the arrow. So for instance, we can define the function F above in such a way as follows:

X is a set of all real numbers, $Y = \{F(x)|\, x \in X\}$, $F: x \longrightarrow 2x + 1$, and $y = F(x)$.

And we can put it this way, too:

X is a set of all real numbers, $Y = \{F(x)|\, x \in X\}$, $x \xrightarrow{F} 2x + 1$, and $y = F(x)$.

That is, $(F: x \longrightarrow 2x + 1) \equiv (x \xrightarrow{F} 2x + 1)$.

So from the definition of F above, too, we can see X is the domain, and Y is the range, and for each value of x, the output gets generated by the expression $2x + 1$, and therefore, we can set: $y = F(x) = 2x + 1$ for x real, or just $y = F(x) = 2x + 1$, because in this case, it is assumed the domain is a set of all real numbers.

And there can be many other ways we can define a function.
For instance, we can sometimes use parameters, too.
And such a parameter can be called a medium.

Using a parameter to make a function, we parameterize a variable to produce a function.

Then, the variable is the output variable in the function produced, and the parameter becomes the input variable.

So the function is a function of the parameter, and is called a parametric function.

Assuming for instance, $x = t + 1$ when $t > 0$, and parameterizing x to make a function called f, we can get: $x = f(t) = t + 1$ for $t > 0$.

Then, we can say x is parameterized, and t is the parameter, and f is called a parametric function, where x is the output variable, so f is a function of t, which is thus, the input variable.

Parameterizing a variable to produce a function, we say we do parameterization. Doing parameterization though, we usually parameterize two or more variables. So we get two or more functions that are parametric.

And we use the same parameter as the input variable of each parametric function. So the parameter can be called the common input variable of the functions. That is, the parameter goes between the functions. And thus, we can connect the functions by means of the parameter.

That is, all the output variables get connected through the medium called the parameter, and then, a new function or an equation gets produced.

In other words, we get a connective expression between the output variables by means of the parameter, and the expression is the expression of the new function or the equation we get.

> Suppose for instance, $x = f(t) = t + 1$ for $t > 0$, and $y = g(t) = t^2$ for $t > 0$.
>
> Then, f and g both are functions of t, and can be called parametric functions, and x and y are output variables, so connecting the two output variables, we can come up with another function.
>
> To begin with, from $x = f(t) = t + 1$, we can get: $t = x - 1$ for $t > 0$.
>
> So we get: $t = x - 1 > 0 \Rightarrow x > 1$, and thus, we get: $t = x - 1$ for $x > 1$.
>
> And we have: $y = g(t) = t^2$, too, so we get: $y = t^2 = (x - 1)^2 \Rightarrow y = (x - 1)^2$ for $x > 1$.
>
> Naming the function above, $h(x)$, we get a function $y = h(x) = (x - 1)^2$ for $x > 1$.

Connecting functions through a parameter, we say we do a parametric transformation, which is covered in the book, **GRAPH OPERATIONS**.

Note:

We can choose any letter for a variable insofar as the consistency is maintained. In other words, it doesn't matter what letter we take as the variable as long as the consistency is maintained.

So we can freely choose any letter provided we keep the consistency. Thus, for instance, "$y = g(x) = 8(3 - x)$ where $x \geq 1$." is the same as "$t = h(s) = 8(3 - s)$ where $s \geq 1$."

That is, the two functions $g(x)$ and $h(s)$ are the same functions.

In other words, though both have different names, both are the same, because both have the same domain, and the same expression.

The input variable s is no more than a container for one input value at a time, and covers all the values in the domain of the function g. And the output variable y is a container for one output value at a time, and covers all the values in the range.

By the way, defining a function, we define a particular function. Defining a function particular, we make a function. Making a function, we can say we make a function definition, too.

And making such definitions, we can have two choices. One is a full definition, and the other is a short definition. It depends on how much detail we show on the function we define. And also, we can call a short definition a brief definition or a semi-definition, too.

A full definition is composed of the name, the domain and the expression. Quite often though, the expression is missing. Then, we can call it a short definition.

If the expression is unavailable or unnecessary for the moment, we can make a short definition using the name and domain only, and can call it the minimal definition.

So for instance, a function definition can be: $y = f(x)$ for $x > 2$, or $t = g(s)$ for s real. And if we get: $t = g(s)$ only, it is assumed that the domain is a set of all real numbers.

3. How do we know if it is one-to-one?

Sometimes, we need to make sure if a function is one-to-one. Then, we need to give it a test to see if it produces a different output for each of all the inputs.

We cannot do so however, for each and every input, since the domain is usually an interval in a number line, and thus, has infinite inputs. So no matter how many numbers we may pick from the domain, the test is insufficient. How then, can we do the test?

We cannot just pick several numbers at random from the domain and do the test either, because it doesn't confirm every case, in other words, the fact that a different output gets produced for each of all the inputs. What else then, can we do?

In *math*, *arbitrary* means not just *random* but *comprehensive*, too.
So a value arbitrary in math has such generality, and can cover all the values in a particular kind, or a category. So it can cover an interval in a number line, too.

What then, do we need to work with doing such a test comprehensively and at random?

We have a tool that can cover all the values in an interval in a number line as the *x*-axis. What tool then, can it be?

It is a constant. A constant can cover all values (numbers) in a kind or a category, and thus, can cover an interval, too, in a number line, or can cover even an entire number line itself. It can cover one value at a time, though.

So what can we do with a constant?

We can use a constant as an input.   How?

We need to use two constants, and let them be different, and belong to the domain so that each can cover the domain.   Why two, though?

Showing that two outputs are different for the two constants, which are two inputs, we show that the function is one-to-one.   How come?

Since we have used constants that can cover all the values applicable, we have shown that no matter what two different inputs we may use, we get two different outputs, so it is a one-to-one function.

How can we show that the two outputs are different, though?

Suppose A and B are the same, that is, $A = B$. Then, we get: $A/B = 1$ if $B \neq 0$, of course, or $A - B = 0$, where A and B both can be 0, of course.

In other words, if $A \neq B$, we get: $A/B \neq 1$ if $B \neq 0$, of course, or $A - B \neq 0$.

(Though it's quite natural that no division by 0 is allowed, the author still makes mistakes a lot, together with lots of other mistakes, too, of course.)

• Assuming for instance, $f(x) = 2x + 1$, let's see now, if the function f is one-to-one.

Suppose first, u and v are constants, and $u \neq v$.

Then, we get: $f(u) - f(v) = (2u + 1) - (2v + 1) = 2(u - v)$.

Next, we have: $u \neq v$, so we get: $u - v \neq 0$.

Thus, we get: $f(u) - f(v) \neq 0 \Rightarrow f(u) \neq f(v)$, so f is one-to-one.

• Assuming next, for another instance, $f(x) = 2x^2 + 1$ for $x \geq 0$, let's check to see if the function f is one-to-one.

Suppose first, u and v are constants, u and v both ≥ 0, and $u \neq v$.

Then, we get: $f(u) - f(v) = (2u^2 + 1) - (2v^2 + 1) = 2(u^2 - v^2) = 2(u + v)(u - v)$.

Next, we have: $u \neq v$, so we get: $u - v \neq 0$.

And we have: u and v both ≥ 0, too, together with $u \neq v$.

So we get: $u \neq -v$, because u and v both cannot be 0 concurrently, and are not negative. That is, we get: $u + v \neq 0$.

So in sum, we get: $u - v \neq 0$, and $u + v \neq 0$.

Thus, we get: $(u + v)(u - v) \neq 0 \Rightarrow f(u) - f(v) \neq 0 \Rightarrow f(u) \neq f(v)$, so f is one-to-one.

• Let's this time, suppose a and b are constant, $a > 0$, and $f(x) = ax^2 + b$ for $x \geq 0$, and check to see if the function f is one-to-one.

Suppose u and v are constants, u and v both ≥ 0, and $u \neq v$.

Then, first, $f(u) - f(v) = (au^2 + b) - (av^2 + b) = a(u^2 - v^2) = a(u + v)(u - v)$.

76

Next, we have: $a > 0$, and $u \neq v$, so we get: $a \neq 0$, and $u - v \neq 0$.

And we have: u and v both ≥ 0, too, together with $u \neq v$.

So we get: $u \neq -v$ because u and v both cannot be 0 concurrently, and are nonnegative.

Thus, we get: $u + v \neq 0$. So we get: $f(u) - f(v) \neq 0 \Rightarrow f(u) \neq f(v)$, and thus, f is one-to-one.

Not quite sure?

Checking to see if a function is one-to-one, we want to check to see if the same output cannot be produced for more than one input. And the check has to be for all the inputs.

So we need to make two different inputs arbitrary, put each into the function, and then, check to see if the outputs are different from each other.

Then, first, we can use two constants as the two inputs.
And next, we've got to do some algebra to show that the outputs are different.

So suppose first, u and v are constants, u and v both ≥ 0, and $u \neq v$.
It's because the domain is: $x \geq 0$, and we want to use two different inputs.

Then, using u and v as inputs, we need to show that $f(u) \neq f(v)$.

Then, we can check to see if $f(u) - f(v) \neq 0$ or $\frac{f(u)}{f(v)} \neq 1$, where $f(v) \neq 0$, of course.

Checking to see if $f(u) - f(v) \neq 0$, we get first:

$$f(u) - f(v) = (au^2 + b) - (av^2 + b) = a(u^2 - v^2) = a(u + v)(u - v).$$

Then, we want to show: $u + v \neq 0$. Why not show $a(u + v)(u - v) \neq 0$, though?

That's because:
We have: $a > 0$, which means $a \neq 0$, too. And also, we have: $u \neq v$, so we get: $u - v \neq 0$.

Thus, we get: $a(u - v) \neq 0$. So next, showing: $u + v \neq 0$, what can we use?

We have: u and v both ≥ 0. So both u and v are not negative.

And we have: $u \neq v$, too.

Thus, we can see that u and v both cannot be 0 at the same time.

So we get: $u \neq -v$ because u and v both cannot be 0 concurrently, and are not negative.

(And for the same reason, we get: $-u \neq v$, too, which is the same as the one above, though.)

Thus, we get $u + v \neq 0 \Rightarrow a(u + v)(u - v)$, and in turn, we get: $f(u) - f(v) \neq 0$.

So we get: $f(u) \neq f(v)$, and thus, f is a one-to-one function.

And putting all the ideas in sum, we can put it the way below:

Suppose u and v are constants, u and v both ≥ 0, and $u \neq v$.

Then, first, $f(u) - f(v) = (au^2 + b) - (av^2 + b) = a(u^2 - v^2) = a(u + v)(u - v)$.

Next, we have: $a > 0$, and $u \neq v$, so we get: $a \neq 0$, and $u - v \neq 0$.

And we have: u and v both ≥ 0, too, together with $u \neq v$.

So we get: $u \neq -v$ because u and v both cannot be 0 concurrently, and are not negative.

Thus, we get: $u + v \neq 0 \Rightarrow f(u) - f(v) \neq 0 \Rightarrow f(u) \neq f(v)$, so f is one-to-one.

• Let's now, for another instance, assume: $g(x) = (x - 1)(x - 2)(x - 3)$, and check to see if g is a one-to-one function.

We can get: $g(1) = (1 - 1)(1 - 2)(1 - 3) = 0$ and $g(2) = (2 - 1)(2 - 2)(2 - 3) = 0$.

So we get: $g(1) = g(2)$. Therefore, the function g is not one-to-one.

How come can we just produce one example only?

This time, we have a little different situation.

It is not always the case where we have to show a function is one-to-one.
That is to say that if we can, we can also show that it is not the case.

Showing it's not the case, we can produce a particular example, called a *counterexample*, which can be called a counter evidence, too, of course.
Producing a counterexample, we have only to show one example where we can see it is not the case. So one counterexample only is enough.

What specifically though, do we need to show in such a counterexample?

We need to show that the function g produces the same output for two different inputs. In this case though, we don't have to make inputs arbitrary. That is, we can use as inputs some numbers that can cause the same output so that they can make a counterexample.

So coming up with two inputs, and showing the outputs are the same, we have produced such a counterexample.

Now of the function g, the domain is a set of all real numbers.
And we know that 1 and 2 are real numbers, and that $1 \neq 2$.

And we can get: $g(1) = (1 - 1)(1 - 2)(1 - 3) = 0$ and $g(2) = (2 - 1)(2 - 2)(2 - 3) = 0$.

So we get: $g(1) = g(2)$. Therefore, the function g is not one-to-one.

4.0. Composite Functions 1

Using a function, we can connect data sets. And basically, two sets get connected, one is the domain, and the other is the range. Sometimes though, we make a function putting together two functions, too. Putting the two together in fact, we combine the two.

And combining the two, we put one into the other. Or rather, we blend the two together. Then, we can get a new function. And the new function is called a composite function.

A composite function is a function, too. So we can make it blend into another function, and can get another composite function. So after all, making a function particular, we can blend together not just two but thee or more functions, too. And the particular function is a blend of functions, and is called a composite function.

So making a composite function, we can say we blend or mix functions together. And thus, it can be called a blend or mix of functions.

As mentioned above however, blending more than two functions, we don't just mix them all at once.
We mix two at a time, and not three or more at once.
Making a composite function mixing together more than two functions, we mix them sequential manner.

We put them in a sequence, and then, begin with the first two. Then, we can get the blend of the two, and the blend is a composite function.
Next, mixing the composite function into the third one in the sequence, we can get another composite function.
And so forth until the last function, so eventually, we can get a mix of all the functions in the sequence.

Mixing two functions though, we don't just mix the two together.
We mix one into the other.
We cannot however, just mix one into the other.   Why not?

We know a function has its domain and range.
And so to speak, taking a composite of two functions, we put the range of one into the domain of the other, if you will.
And thus, mixing functions together, we need to be careful with the sizes of the domains and the ranges.   Whose domain and whose range though?

The range of the function that blends into. And the domain of the other function.

The range of the function that blends into has to be a subset of the domain of the other.
(A subset is a set smaller or equal.)
If the range is bigger, that is, the domain is smaller, they don't mix. Why not?

We will get to see why not shortly.   Let's see first, how blending works.

Blending two functions, we plug one into the other. Then, we get another function, called a composite function.

> And blending, we blend the expression of one function into the expression of the other.
> • In short, blending, we put one function into the other.   How?

> Blending them, we plug the expression of one function into the input variable in the expression of the other function. More specifically, each and every input variable in the expression of the other function gets the expression.   How?

Suppose for instance, a function $f(x)$ blends into another function $g(x)$, $2x$ is the expression of $f(x)$, and $x^2 + x$ is the expression of the other function $g(x)$.

Then, after blending, we get a new function where the expression is:
$(2x)^2 + (2x) = 4x^2 + 2x$.   How come, though?

What does the expression of a function stand for?

The expression of a function produces each and every output of the function. So it can be said to stand for all the outputs of the function.

And the input variable in a function gets each and every input of the function. So it can be said to stand for all the inputs in the function.

So in sum, an expression stands for a range, and an input variable stands for a domain. Or rather, an expression is no other than a range, and an input variable is no other than a domain.

And thus, blending the expression of $f(x)$ into the expression of $g(x)$, we put the expression of f into the input variable, that is, x in $g(x)$.

  • So the expression of f gets into each and every input variable in the expression of g.

So for instance, blending $x + 1$ into $x^2 + 2x$, we get: $(x + 1)^2 + 2(x + 1) = x^2 + 4x + 3$.

What can we notice then, from the fact above?

We know the expression of a function produces each and every output of the function.

So we can notice that each output of the function that blends into is used as each input of the other function.
That is, every output of the function that blends into is every input of the other function.

  • In other words, the range of the function blends into is the domain of the other.

Suppose now, blending *f* into *g*, we get a composite function **h**.
What then, is the domain of the composite function **h**?

We know the range of *f* is used as the domain of *g* during the process of making **h**.

And of course, the range of *f* is made by the expression of *f* using the domain of *f*.

That's because the expression gets inputs from the domain, and produces outputs, which belong to the range.

So the domain of the composite function **h** is the domain of *f*, the function blends into.

And putting the ideas above into a diagram, we can get:

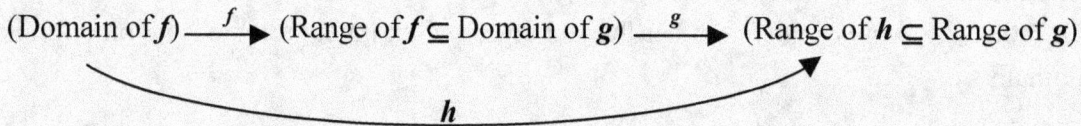

So we get: (Domain of *f*) $\xrightarrow{h}$ (Range of **h**).   And Range of **h** $\subseteq$ Range of **g**.

So the domain of *f* is the domain of **h**, and the range of **h** is a subset of the range of **g**.

What then, can be a problem when we make a composite function using two functions?

Making a composite function, we cannot just mix any functions, because the functions have to fit. Or rather, the number sets have to fit.

  • The range of the function blends into has to fit the domain of the other function.

That's because the range of the one blends into will be the domain of the other.

And the domain of the function blends into is the domain of the composite function.
And we will see shortly more concrete details on how it is the case.

Now, the range of the one blends into has to be a part of or the same as the domain of the other. So anyway, it is ideal that such a range is the same as such a domain.

We know domains and ranges are sets (of numbers). So we want to make sure that the range of the one blends into is a subset of the domain of the other.

(If A is a subset of a set B, we get: $A \subseteq B$, which means A is either B or a part of B.)

Let's now make a case where the range of f is not a subset of the domain of g, and see what will happen.

Suppose for instance, the domain of g is a set of all positive numbers, the range of f is a set of all positive numbers and 0, that is, a set of all numbers ≥ 0, and we want to make f blend into g to get a composite function called h.

Then, the range of f has to be the domain of g, but 0 cannot be an input of g, because g is not defined for 0, since 0 is not in the domain of g. So if in g, 0 is used as an input, g crashes, if you will. What problem then, will the composite function h get?

The function h cannot hold for _some number_, which is of course, in the domain of h. How come?

We know 0 is in the range of f, so 0 is an output for _some number_ in the domain of f.

And we know the domain of f is the domain of the composite function h.

So the _some number_ the function h cannot hold for is the _some number_ in the domain of f.
And thus, h will crash for the _some number_.

So the function f cannot blend into the other function g.
That is to say that we cannot get the composite function h.

Let's now make a bit more concrete case where we can see better how a composite function gets made.

Suppose for instance, mixing two functions A and B, we make a new function C. Suppose specifically, we put A into B to get C.

Then, C is a composite function where A blends into B.
And one thing for sure now is that the domain of C is the domain of A.

Suppose now, X_A is the domain of A, Y_A is the range of A, X_B is the domain of B, Y_B is the range of B, X_C is the domain of C, and Y_C is the range of C.

So X stands for a domain, and Y stands for a range. And A and B can be set as follows.

$$X_A \xrightarrow{A} Y_A \qquad X_B \xrightarrow{B} Y_B$$

Then first, putting A into B, we need to have: $Y_A \subseteq X_B$, which means Y_A is a subset of X_B.

That is, the range of A is equal to or a part of the domain of B.

• So the mixing process goes as follows: $X_A \xrightarrow{A} (Y_A \subseteq X_B) \xrightarrow{B} Y_B$

Thus, all the inputs to be used in B are all the outputs made in A.

That is, the range of A will be the domain of B. So the mixing process continues as below, and then ends up with the function C, which is the composite function.

$$X_A \xrightarrow{A} Y_A \xrightarrow{B} Y_C \Rightarrow X_C \xrightarrow{C} Y_C \quad \text{In } C, \text{ we have: } X_C = X_A, \text{ of course.}$$

And thus, all the inputs of C are from X_A, that is, the domain of C is the domain of A. What then, about the range?

The range of C can be the range of B. What do we mean by 'can be' though?

It means that the range of C can be other than the range of B, too.
In other words, it is not always the case where $Y_C = Y_B$.

That's because the range of A can be a subset of the domain of B, and thus, can be smaller than the domain of B.

And in that case, it can also be the case the range of C can be a subset of the range of B.

4.1. Composite Functions 2

(Note that this section is for you, only if you really have to look at composite functions very closely.)

What if we want to get a new function combining three functions?

Two of the three get combined first, then a new function gets made, and is of course, a composite function.

And next, the new function gets put into the other, then another new function gets made, and is the composite function we want. So in the composite function, the domain is the domain of one of the two functions combined first.

Suppose for instance, mixing three functions A, B, and C, we make a new function F. Suppose also, we begin with mixing A into B.

Then first, putting A into B, we get a new function called D, for instance.

So D is a composite function, where A blends into B.

In turn, putting the function D into the function C, we get another new function, which is the function F stated above.

In short, putting A into B, we get D, then putting D into C, we get F.

Suppose now again, X_A is the domain of A, Y_A is the range of A, X_B is the domain of B, Y_B is the range of B, X_C is the domain of C, and Y_C is the range of C.

$$X_A \xrightarrow{A} Y_A \qquad X_B \xrightarrow{B} Y_B \qquad X_C \xrightarrow{C} Y_C$$

Then first, putting A into B, we need to have: $Y_A \subseteq X_B$. That is, Y_A is a subset of X_B.

In other words, the range of A is equal to or a part of the domain of B.

$$X_A \xrightarrow{A} (Y_A \subseteq X_B) \xrightarrow{B} Y_B \qquad X_C \xrightarrow{C} Y_C$$

Then, we get a new function D, which is a composite function where A blends into B.

$$X_A \xrightarrow{A} Y_A \xrightarrow{B} Y_D \underset{D}{\frown} \Rightarrow X_D \xrightarrow{D} Y_D \quad \text{In } D, \text{ we have: } X_D = X_A, \text{ of course.}$$

Thus, the domain of D is the domain of A. So D is a function from the domain of A to the range of D, which can be the range of B, too. And therefore, we get: $X_D = X_A$, and it can be the case where $Y_D = Y_B$. And next, we need to blend D into C.

$$X_A \xrightarrow{D} Y_D \qquad X_C \xrightarrow{C} Y_C$$

So putting D into C, we need to have: $Y_D \subseteq X_C$.

In other words, the range of D is a subset of the domain of C.

$$X_A \xrightarrow{D} (Y_D \subseteq X_C) \xrightarrow{C} Y_C$$

Then, we get the new function F, which is a composite function where D blends into C.

$$X_A \xrightarrow{D} Y_D \xrightarrow{C} Y_F \underset{F}{\frown} \Rightarrow X_F \xrightarrow{F} Y_F \quad \text{In } F, \text{ we have: } X_F = X_A, \text{ of course.}$$

Thus, the domain of F is the domain of A. So F is a function from the domain of A to the range of F, which can be the range of C, too. And therefore, we get: $X_F = X_A$, and it can be the case where $Y_F = Y_C$.

Now, we know that the function D is a composite function where A blends into B, and that the function F is a composite function where D blends into C.

So the function F is the composite function where all the three functions A, B, and C blended together, and thus, in short, is a blend of A, B, and C.

So we can make a new function blending two functions.
Blending the two functions though, we blend one into the other.
And the range of the function blends into is a subset of the domain of the other.

Then, we get a new function, which is called a composite function, which is a blend of two functions.

And repeating the same process with the composite function made and another function, we can get another composite function, which is a blend of three functions.

And of course, the range of the composite function made earlier is a subset of the domain of the other function. Also, the domain of the composite function newly made is the domain of the composite function made earlier.

So blending two functions the way above, we can get a composite function.

How then, do we actually do such blending to make a composite function?

Defining a function particular, we make a function.
That is, making a function definition, we make a function.
Defining a composite function therefore, we make a composite function.

So let's make one now. That is, we are going to make a definition of a composite function.

A composite function is a mix of two or more function, so such a definition begins with at least two function definitions. So for instance, it can go as follows:

X_f, Y_f, X_g, and Y_g are sets of real numbers, where $Y_f \subseteq X_g$,

$f: X_f \longrightarrow Y_f$, $g: X_g \longrightarrow Y_g$, $y = f(x)$, $z = g(y)$, and $z = h(x)$.

90

(Note that a set of real numbers is a set of not all but some real numbers. So X_f, Y_f, X_g, and Y_g can be different sets.)

The definition above is saying that h is a composite function where f blends into g. How come?

Let's not take a closer look. So let's break apart a composite function, and see in detail how it gets made, and of course, how it works, too.

> We have:
>
> X_f, Y_f, X_g, and Y_g are sets of real numbers, where $Y_f \subseteq X_g$,
>
> $f: X_f \longrightarrow Y_f$, $g: X_g \longrightarrow Y_g$, $y = f(x)$, $z = g(y)$, and $z = h(x)$.

First, since Y_f is the range of f, all the numbers in Y_f are all the outputs of f.

And next, once f has blended into g, we get: $X_g = Y_f$, that is, the domain of g changes to the range of f, that is, Y_f, so all the outputs of f are all the inputs of g.

More specifically, each time an output is made in the function f, the output is an input of the function g. So we can put it the way below:

We know y gets every output of f.
Also, since $y = f(x)$, $f(x)$ gets each output of f, too.

Next, we have: $z = g(y)$, so the expression of g is an expression in terms of y as $2y + 1$.
So for instance, we can have: $z = g(y) = 2y + 1$ for $y \in Y_f$.

And we have: $y = f(x)$, too, so we can set: $z = g(f(x))$ since $z = g(y)$.
That is, setting: $z = g(f(x))$, read as g of f of x, we put $f(x)$ into y in $z = g(y)$.

And $g(f(x))$ means that each input of g is each output of f.

And since $y = f(x)$, the expression of f is an expression in terms of x as $x + 1$.

So for instance, we can have: $y = f(x) = x + 1$ for $x \in X_f$.

Then, setting: $z = g(f(x))$, we can put $(x + 1)$ into y in $z = g(y)$, since $f(x) = x + 1$.

So we get: $z = g(y) = 2y + 1 \Rightarrow z = 2y + 1 = 2(x + 1) + 1 = 2x + 3 \Rightarrow z = 2x + 3$, which is an expression in terms of x, and can be used as an expression of a function.

Now, we have: $z = h(x)$, which indicates the expression of h is expressed in terms of x.

And we have: $z = 2x + 3$, too. So we can see that $z = h(x) = 2x + 3$.

And we know x gets its value in the domain of $f(x)$, and thus, so does x in $z = h(x)$.

That is, x in $z = h(x)$ gets its value in the domain of f, too, because x was from $(x + 1)$, which is the expression of f. So the domain of h is the domain of f, and thus, is X_f.

So putting threads together, we can say that h is a function that has the domain of f.

And thus, h is a composite function where f blends into g, and is as follows:

$z = h(x) = 2x + 3$ for $x \in X_f$.

And of course, $X_h = X_f$.   And we can put h the way below, too:

$h: X_f \longrightarrow Y_h, y = f(x) = x + 1, z = g(y) = 2y + 1$, and $z = h(x)$.

Now, you may want to take a little break, and then go back to the beginning of this example, and go over the entire processes. A function is not a fair game, much less a composite function.

And thus, what's happening in h is as good as what's happening sequentially in f and g.

It begins with f, of course, which is the function blends into.   How come?

Example is the best teacher, isn't it? So let's get back to the example. We have:

$z = h(x) = 2x + 3$ for $x \in X_f$.

$y = f(x) = x + 1$ for $x \in X_f$, and *after blending*, we get: $z = g(y) = 2y + 1$ for $y \in Y_f$.

So to begin with, suppose in h, x gets 1 as an input from the domain X_f.

Then, z gets the output, which is: $2 \cdot 1 + 3 = 5$, which belongs to the range of h, of course.

And suppose next, in f, x gets 1 as an input from its domain, which is X_f.

Then, in f, y gets the output, which is: $1 + 1 = 2$, which belongs to the range Y_f.

And in turn, the output 2 is used as an input of g, since the domain of g is the range of f.

Thus, in $z = g(y)$, y gets 2, the output just has been made in f, so z gets the output for 2, the value y has, and the output is: $2 \cdot 2 + 1 = 5$, which is an element in the range of h.

So the major blending happens the way below:

The expression of f blends into the expression of g to result in the expression of h.

That is, having $(y = x + 1)$ blend into $z = 2y + 1$, we get: $z = 2(x + 1) + 1 = 2x + 3$.

So we can put a bit differently, the definition of the composite function h the way below:

X_f, Y_f, X_g, and Y_g are sets of real numbers, where $Y_f \subseteq X_g$,
$f: X_f \longrightarrow Y_f$, $g: X_g \longrightarrow Y_g$, $y = f(x)$, $y = g(x)$, and $y = h(x) = g\{f(x)\}$.

The definition above is of course, no other than the definition stated earlier, which is:

X_f, Y_f, X_g, and Y_g are sets of real numbers, where $Y_f \subseteq X_g$,
$f: X_f \longrightarrow Y_f$, $g: X_g \longrightarrow Y_g$, $y = f(x)$, $z = g(y)$, and $z = h(x)$.

Both look quite different though, because the variables are different, and we don't even see z in the first of the two definitions above.   How come then, both are the same?

In functions, what matters regarding variables is not only their names but their roles, too.

So we want to make sure: *which has what value in kind*.

And thus, we can put the definition above the way below, too:

X_f, Y_f, X_g, and Y_g are sets of real numbers, where $Y_f \subseteq X_g$,
$f: X_f \longrightarrow Y_f$, $g: X_g \longrightarrow Y_g$, $y = f(x)$, $v = g(u)$, and $t = h(x) = g\{f(x)\}$.

What's important regarding variables in the definition above is made of two facts.

• One is that *after blending*, u gets values from Y_f, which is the range of f, so when blending, we can set: $u = f(x)$, and therefore, we get: $g\{f(x)\}$, which is $g(f)$, for short.

So what $g(f)$ is saying is that we put the expression of f into every u in the expression of g.

• And the other fact is that in $h(x)$ and $f(x)$ both, the variable x gets values from X_f, which is the domain of $f(x)$. So h and f share the same domain, which is the domain of f.

That's because in $h(x)$, x is from $f(x)$, which is the function blends into. And the first fact above explains it.

So we can try blending $f(x)$ into $g(u)$ to get the composite function $h(x)$. Not quite clear?

A function is not a fair game, still less a composite function.
So let's see first, what's happening in the blending processes in the definition below:

X_f, Y_f, X_g, and Y_g are sets of real numbers, where $Y_f \subseteq X_g$,
$f: X_f \longrightarrow Y_f$, $g: X_g \longrightarrow Y_g$, $y = f(x)$, $y = g(x)$, and $y = h(x) = g\{f(x)\}$.

94

To begin with, we had: $y = f(x) = x + 1$ for $x \in X_f$, and $z = g(y) = 2y + 1$ for $y \in Y_f$.

And we now have: $y = f(x)$ for $x \in X_f$, and *after blending*, $y = g(x)$ for $x \in Y_f$.

So we get: $y = f(x) = x + 1$ for $x \in X_f$, and $y = g(x) = 2x + 1$ for $x \in Y_f$.

And that's not it, we have: $y = h(x) = g\{f(x)\}$, too. And this is the place where the major blending begins.   So the question is, what do we mean by $g\{f(x)\}$?

Putting $f(x)$ into x in $g(x)$, we get: $g(f(x))$, and making it more legible, we can put it this way: $g\{f(x)\}$. And $g\{f(x)\}$ means that each input of g is each output of f.

Now, we have: $y = f(x) = x + 1$, and $y = g(x) = 2x + 1$.

So putting $f(x)$ into x in $y = g(x)$, we get: $y = g\{f(x)\} = 2f(x) + 1$.

And we have: $f(x) = x + 1$, so we get: $y = 2f(x) + 1 = 2(x + 1) + 1 = 2x + 3$, which is an expression in terms of x, but is certainly not the expression of the function $y = g(x)$. What expression then, is it?

We have not only $y = g(x)$. But we have: $y = h(x)$, too.

So we get: $y = h(x) = 2x + 3$ for $x \in X_f$.   How come $x \in X_f$, though?

That's because x is from $(x + 1)$, which is the expression of f, of which the domain is X_f.

Now, let's see next, what's happening in the blending procedure in the definition below:

X_f, Y_f, X_g, and Y_g are sets of real numbers, where $Y_f \subseteq X_g$,

$f: X_f \longrightarrow Y_f$, $g: X_g \longrightarrow Y_g$, $y = f(x)$, $v = g(u)$, and $t = h(x) = g\{f(x)\}$.

To begin with, we had:

$y = f(x) = x + 1$ for $x \in X_f$, and $z = g(y) = 2y + 1$ for $y \in Y_f$.

And we now have: $y = f(x)$ for $x \in X_f$, and *after blending*, $v = g(u)$ for $u \in Y_f$, so we get:

$y = f(x) = x + 1$ for $x \in X_f$, and $v = g(u) = 2u + 1$ for $u \in Y_f$.

And of course, we have: $t = h(x) = g\{f(x)\}$, too.
So again, this is the place where the major blending happens.

So the question is again, what do we mean by $g\{f(x)\}$?

Putting $f(x)$ into u in $g(u)$, we get: $g\{f(x)\}$, and $g\{f(x)\}$ means that each input of g is each output of f.

Now, we have: $y = f(x) = x + 1$, and $v = g(u) = 2u + 1$.

So putting $f(x)$ into u in $v = g(u)$, we get: $v = g\{f(x)\} = 2f(x) + 1$.

And we have: $f(x) = x + 1$, so we get: $v = 2f(x) + 1 = 2(x + 1) + 1 = 2x + 3$, which is an expression in terms of x, which is clearly not the expression of the function $v = g(u)$.

What expression is it, then?

We have: $t = h(x)$, too. So we get: $t = h(x) = 2x + 3$ for $x \in X_f$.

And x in $h(x)$ is from $f(x)$, of which the domain is X_f, so in $h(x)$, we have: $x \in X_f$.

And t gets outputs of the function h, and thus, is just the output variable in h.

What if we like to change to w, for instance, the name of the input variable in $f(x)$?

We can do so, and yet, we want to change the name of the input variable in the function $h(x)$, too. Then, the definition above simply changes its look, and will be the one below:

X_f, Y_f, X_g, and Y_g are sets of real numbers, where $Y_f \subseteq X_g$,

$f: X_f \longrightarrow Y_f$, $g: X_g \longrightarrow Y_g$, $y = f(w)$, $t = g(u)$, and $t = h(w) = g\{f(w)\}$.

So the name of the input variable in the composite function is the same as that of the input variable in the function blends into.

That's because getting the expression of the composite function, we put into each and every input variable in the expression of the other function, the expression of the function blends into.

In short, the input variable of a composite function is that of the function that blends into.

That is, both names have to be the same.

So a composite function is a blend of functions.

It depends on however, the way functions get blended together.
That is, the blending sequence matters.

4.2. **Composite Functions 3**

(Note again that this section is for you, only if you really have to look at composite functions very closely.)

How then, can we specify the blending sequence making a composite function?

Defining a composite function, we have another way of doing it.

We can use an operator as + or – when making a composite function.

As the operator, we use a dot •, which is normally a bit bigger than the tiny one · often used in a multiplication as in $5 \cdot 3 = 15$. We can call it a composite-function-operator.

> So for instance, a composite function $g\{f(x)\}$ can be put in $g \bullet f(x)$, which is simply read as: g dot f of x. And thus, we can set: $g \bullet f(x) = g\{f(x)\}$.
>
> Or briefly, we can put it this way, too: $g \bullet f = g(f)$.

So we can use such an operator making a definition of a composite function.

And for instance, we can define a composite function in such a way as follows:

X_f, Y_f, X_g, Y_g, and W are sets of real numbers, where $Y_f \subseteq X_g$,

$f \colon X_f \longrightarrow Y_f$, $g \colon X_g \longrightarrow Y_g$, and $g \bullet f \colon X_f \longrightarrow W$.

(And of course, it can be the case where $W = Y_g$.)

A bit more specifically, we can put it the way below, too:

$X_f, Y_f, X_g,$ and Y_g are sets of real numbers, where $Y_f \subseteq X_g$,

$f: X_f \longrightarrow Y_f, g: X_g \longrightarrow Y_g, y = f(x)$, and $g \bullet f : x \longrightarrow g\{f(x)\}$ or just $g(f)$.

And we can notice that $g \bullet f$ can be taken for a function designator, that is, a name.

So specifying the input variable in the composite function $g \bullet f$, we can put it this way, too: $(g \bullet f)(x)$, which is however, no other than $g \bullet f(x)$.

It's a bit bulky though, to carry around with, so we often give a single name to it, and for instance, we can set: $h = g \bullet f$. So in that case, we can also set: $h(x) = (g \bullet f)(x)$.

And for instance, we can define a composite function in such a way as follows:

$X_f, Y_f, X_g, Y_g,$ and Y_h are sets of real numbers, where $Y_f \subseteq X_g$,

$f: X_f \longrightarrow Y_f, g: X_g \longrightarrow Y_g, h: X_f \longrightarrow Y_h$, and $h = g \bullet f$.

So we can see that:

$$X_f \xrightarrow{f} Y_f \xrightarrow{g} Y_h \quad \Rightarrow \quad X_f \xrightarrow{g \bullet f} Y_h.$$

And of course, it can be the case where $Y_h = Y_g$.
And it is worth to note that it is usually the case where: $g \bullet f \neq f \bullet g$.

That is because first:

In $g \bullet f$, the function blends into is f, so the domain is the domain of f.

In $f \bullet g$ though, the function blends into is g, so the domain is the domain of g.

And next, even if the domains are identical for both functions, the expressions can be still different. So it is usually the case where $g \bullet f \neq f \bullet g$.

Why usually, though? Can it be then, the case where $g \bullet f = f \bullet g$, too?

Yes, it can. Suppose for instance, $y = f(x) = x$ for x real, and $y = g(x) = x^2 + 1$ for x real.

Then, we get: $g \bullet f(x) = x^2 + 1$, and also, $f \bullet g(x) = x^2 + 1$.   How come?

We have: $f(x) = x$, so we can just set: $f = x$.
And we have: $g \bullet f(x) = g(f)$, and $g(x) = x^2 + 1$, too.

So we get: $g \bullet f(x) = g(f) = f^2 + 1 = x^2 + 1$, since $f = x$. And thus, $g \bullet f(x) = x^2 + 1$ for x real.

And by the same token, we get: $f \bullet g(x) = f(g)$, where $f(x) = x$, and $g = x^2 + 1$.

So we get: $f \bullet g(x) = f(g) = g = x^2 + 1$. And thus, $g \bullet f(x) = x^2 + 1$ for x real.

Now, the function $y = f(x) = x$ is an identity function. So we can notice that if either of the two functions is an identity function, we can get: $g \bullet f = f \bullet g$.

Suppose this time, $y = f(x) = x + 1$, and $y = g(x) = x^2$.

Then, we get: $g \bullet f(x) = g(f) = f^2 = (x + 1)^2$, but we get: $f \bullet g(x) = f(g) = g + 1 = x^2 + 1$.

• Now, let's take a look at how some examples work.

First, assuming $y = f(x) = x + 1$ for $1 \leq x \leq 2$, and $y = g(x) = 2x + 1$ for $2 \leq x \leq 3$, let's find $y = h(x) = (g \bullet f)(x)$.

Since h is a function, too, finding h, we want to get the expression, along with the domain.

Now, to begin with, assuming X is a domain, and Y is a range, we can put it this way:

$$X_f \xrightarrow{\;f\;} Y_f \xrightarrow{\;g\;} Y_h \Rightarrow X_f \xrightarrow{\;g \bullet f\;} Y_h.$$ And of course, in h, $X_h = X_f$, the domain of f.

So the domain of h is: $1 \leq x \leq 2$. And next, we want to check to see if the range of the function blends into is a subset of the domain of the other function. That is, the range of the function f has to be equal to or a part of the domain of the function g.

So checking it, we get: $1 \leq x \leq 2 \Rightarrow 2 \leq x + 1 \leq 3 \Rightarrow 2 \leq y \leq 3$, which is the range of f.

We know domains and ranges are sets of numbers, and the domain of g is: $2 \leq x \leq 3$.

So the range of f is a set of all numbers from 2 and 3, and so is the domain of g.

The range of f is thus, equal to the domain of g, so h can be defined in the domain of f.

Now, we have: $y = f(x) = x + 1$, $y = g(x) = 2x + 1$, and $y = h(x) = (g \bullet f)(x) = g(f)$.

So we get: $h(x) = g(f) = 2f + 1$, since $g(x) = 2x + 1$.

And thus, $h(x) = 2f + 1 = 2(x + 1) + 1 = 2x + 3$, since $f(x) = x + 1 \Rightarrow f = x + 1$.

Next, the domain of h is the domain of f, so the domain is: $1 \leq x \leq 2$.

And we can see the domain has no problem with the expression where $2x + 3$.

And next, let's see if the range of h is the same as the range of g.

We have: $y = g(x) = 2x + 1$ for $2 \leq x \leq 3$, and $y = h(x) = 2x + 3$ for $1 \leq x \leq 2$.

So beginning with the range of g, we get: $2 \leq x \leq 3 \Rightarrow 4 \leq 2x \leq 6 \Rightarrow 5 \leq 2x + 1 \leq 7$.

And next, finding the range of h, we get: $1 \leq x \leq 2 \Rightarrow 2 \leq 2x \leq 4 \Rightarrow 5 \leq 2x + 3 \leq 7$.

So both are the same, and in this case, we can say that the composite function is a function from the domain of the function blends into to the range of the other function.

Let's now put in a graph all the functions f, g, and h.

Fig. 5.0

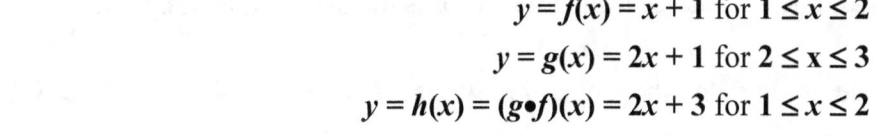

$$y = f(x) = x + 1 \text{ for } 1 \leq x \leq 2$$
$$y = g(x) = 2x + 1 \text{ for } 2 \leq x \leq 3$$
$$y = h(x) = (g \bullet f)(x) = 2x + 3 \text{ for } 1 \leq x \leq 2$$

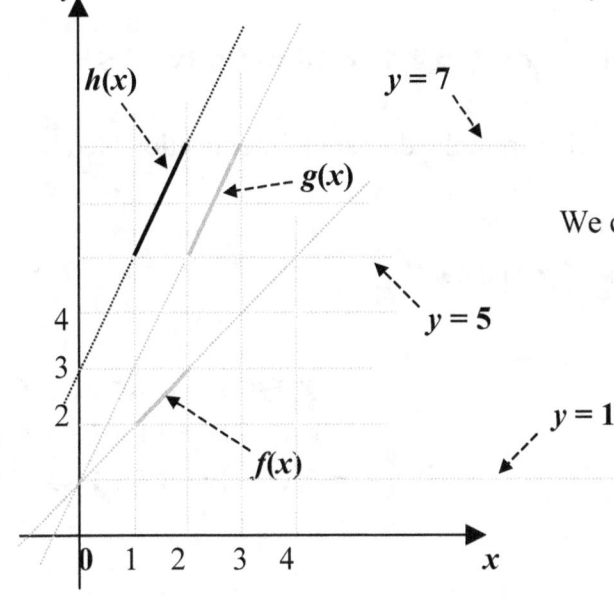

We can actually see h has all the inputs of f and all the outputs of g.

Now, let's for next instance, find $y = h(x) = (g \bullet f)(x)$ assuming $y = f(x) = 2x$ for $1 \leq x \leq 2$, and $y = g(x) = 2x + 1$ for $2 \leq x \leq 4$.   Then first, we may want to put it this way:

$$X_f \xrightarrow{\ f\ } Y_f \xrightarrow{\ g\ } Y_h \quad \Rightarrow \quad X_f \xrightarrow{\ g \bullet f\ } Y_h. \quad \text{And of course, we have: } X_h = X_f.$$

So the domain of h is: $1 \leq x \leq 2$. And next, checking to see if the range of f is a subset of the domain of g, we get: $1 \leq x \leq 2 \Rightarrow 2 \leq 2x \leq 4 \Rightarrow 2 \leq y \leq 4$, which is the range of f.

We know the domain of g is: $2 \leq x \leq 4$.

So the range of f equals the domain of g, and thus, h can be defined in the domain of f.

Now, we have: $y = f(x) = 2x$, $y = g(x) = 2x + 1$, and $y = h(x) = (g \bullet f)(x) = g(f)$.

So we get: $h(x) = g(f) = 2f + 1 = 2(2x) + 1 = 4x + 1$, which has no problem with the domain given. Thus, we get: $y = h(x) = 4x + 1$ for $1 \leq x \leq 2$.

And next, let's see if the range of h is the same as the range of g.

Beginning with the range of g, we get: $2 \leq x \leq 4 \Rightarrow 4 \leq 2x \leq 8 \Rightarrow 5 \leq 2x + 1 \leq 9$.

And next, finding the range of h, we get: $1 \leq x \leq 2 \Rightarrow 4 \leq 4x \leq 8 \Rightarrow 5 \leq 4x + 1 \leq 9$.

So in this case, too, h is a composite function from the domain of the function blends into to the range of the other function.

Let's now put in a graph all the functions f, g, and h.

Fig. 5.1

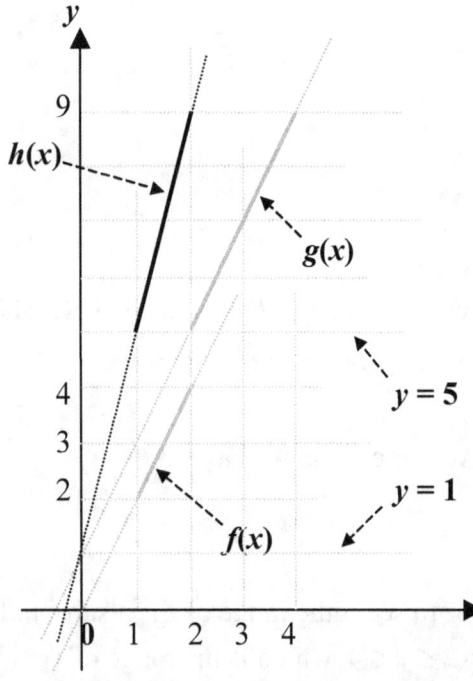

$$y = f(x) = 2x \text{ for } 1 \leq x \leq 2$$
$$y = g(x) = 2x + 1 \text{ for } 2 \leq x \leq 4$$
$$y = h(x) = 4x + 1 \text{ for } 1 \leq x \leq 2$$

In the function h, all the inputs are those of f, and all the outputs are those of g.

Now, let's for next instance, assuming $y = f(x) = x + 1$ for $1 \leq x \leq 2$, and $y = g(x) = x^2$ for $2 \leq x \leq 3$, find $y = h(x) = (g \bullet f)(x)$. To begin with, we can put it this way:

$$X_f \xrightarrow{\ f\ } Y_f \xrightarrow{\ g\ } Y_h \quad \Rightarrow \quad X_f \xrightarrow{\ g \bullet f\ } Y_h.$$ In h, we have: $X_h = X_f$, the domain of f.

So the domain of h is: $1 \leq x \leq 2$. And next, checking to see if the range of f is a subset of the domain of g, we get: $1 \leq x \leq 2 \Rightarrow 2 \leq x + 1 \leq 3 \Rightarrow 2 \leq y \leq 3$, which is the range in f.

We know the domain in g is: $2 \leq x \leq 3$.

So the range in f equals the domain in g, and thus, h can be defined in the domain of f.

Now, we have: $y = f(x) = x + 1$, $y = g(x) = x^2$, and $y = h(x) = (g \bullet f)(x) = g(f)$.

So we get: $h(x) = g(f) = f^2 = (x + 1)^2$, which has no problem with the domain: $1 \leq x \leq 2$.

Thus, we get: $y = h(x) = (x + 1)^2$ for $1 \leq x \leq 2$.   Let's next, check the ranges of h and g.

Beginning with the range of g, we get: $2 \leq x \leq 3 \Rightarrow 4 \leq x^2 \leq 9$.

And next, finding the range of h, we get: $1 \leq x \leq 2 \Rightarrow 2 \leq x + 1 \leq 3 \Rightarrow 4 \leq (x + 1)^2 \leq 9$.

So h is a composite function from the domain of f to the range of g.

Let's now have a look at the graph where all the functions f, g, and h are put.

Fig. 5.2

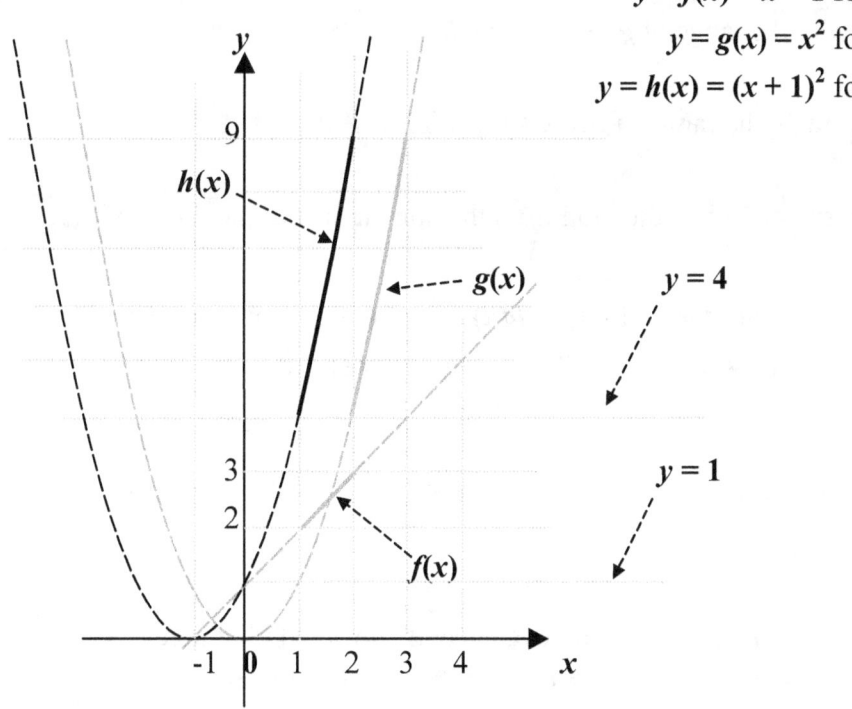

$$y = f(x) = x + 1 \text{ for } 1 \leq x \leq 2$$
$$y = g(x) = x^2 \text{ for } 2 \leq x \leq 3$$
$$y = h(x) = (x + 1)^2 \text{ for } 1 \leq x \leq 2$$

Now, let's for next instance, find $y = h(x) = (g \bullet f)(x)$ assuming $y = f(x) = x^2$ for $x \geq 0$, and $y = g(x) = \sqrt{x}$ for $x \geq 0$. Then first, we can put it this way:

$$X_f \xrightarrow{\;f\;} Y_f \xrightarrow{\;g\;} Y_h \quad \xRightarrow{\;\;} \quad X_f \xrightarrow{\;g \bullet f\;} Y_h \quad \text{In } h, \text{ we have: } X_h = X_f, \text{ the domain of } f.$$

So the domain of h is: $x \geq 0$. And next, checking to see if the range of f is a subset of the domain of g, we get: $x \geq 0 \Rightarrow x^2 \geq 0 \Rightarrow y \geq 0$, which is the range of f.

We know the domain in g is: $x \geq 0$.
So the range of f equals the domain of g, and thus, h can be defined in the domain of f.

Now, we have: $y = f(x) = x^2$, $y = g(x) = \sqrt{x}$, and $y = h(x) = (g \bullet f)(x) = g(f)$.

So we get: $h(x) = g(f) = \sqrt{f} = \sqrt{x^2} = x$, which has no problem with the domain: $x \geq 0$.

Thus, we get: $y = h(x) = x$ for $x \geq 0$. And let's next, check the ranges of h and g.

Beginning with the range of g, we get: $x \geq 0 \Rightarrow \sqrt{x} \geq 0$.

And next, finding the range of h, we simply get: $x \geq 0$.

So in this case, too, h is a function from the domain of f to the range of g.

Now, let's for next instance, find $y = h(x) = (g \bullet f)(x)$ assuming $y = f(x) = x^2$ for $x \geq 0$, and $y = g(x) = \sqrt{x - 4}$ for $x \geq 4$. To begin with, we have:

$$X_f \xrightarrow{\;f\;} Y_f \xrightarrow{\;g\;} Y_h \quad \Rightarrow \quad X_f \xrightarrow{\;g \bullet f\;} Y_h. \quad \text{And of course, we have: } X_h = X_f.$$

So the domain of h is: $x \geq 0$. And next, checking to see if the range of f is a subset of the domain of g, we get: $x \geq 0 \Rightarrow x^2 \geq 0 \Rightarrow y \geq 0$, which is the range of f.

However, the domain of g is: $x \geq 4$.

So the range of f is not a subset of the domain of g, and on the contrary, the domain of g is a part of the range of f. And thus, h cannot be defined in the domain of f.

Let's see though, what will happen in this case.

We have: $y = f(x) = x^2$, $y = g(x) = \sqrt{x-4}$, and $y = h(x) = (g \bullet f)(x) = g(f)$.

So we get: $h(x) = g(f) = \sqrt{f-4} = \sqrt{x^2-4}$, which has however, a lot of problem with the domain of f, which is X_f.

That is because the domain is: $x \geq 0$, but if $x = 1$, for instance, what's inside the square root is -3, which is not allowed.

And therefore, we cannot get a composite function that can hold for the domain of the function blends into. And the function blends into is f, of course, in this case.

Let's see now what can be the domain of h. Finding it, we get:

$x^2 - 4 \geq 0 \Rightarrow (x+2)(x-2) \geq 0 \Rightarrow x \geq 2$ or $x \leq -2$.

So the maximum domain of h can be: $\{x \mid x \geq 2$ or $x \leq -2\}$ or simply: $x \geq 2$ or $x \leq -2$.

What do we mean by the maximum domain, though?

The domain of h can be either of: $x \geq 2$, $x \geq 3$, $x \geq 5$, and such, each of which is smaller than the maximum domain above.

We know both functions f and g can be defined for $x \geq 4$, for instance, and also, the function h can be defined for $x \geq 4$.

So if the domain X_f is: $x \geq 4$, we can get a composite function that can be defined in the domain of the function blends into.

And the composite function is: $y = h(x) = \sqrt{x^2 - 4}$ for $x \in X_f$, which is: $x \geq 4$.

And let's this time, find $y = h(x) = (f \bullet g)(x)$ using the two functions above.

Then first, we have::

$$X_g \xrightarrow{\ g\ } Y_g \xrightarrow{\ f\ } Y_h \quad \Rightarrow \quad X_g \xrightarrow{\ f \bullet g\ } Y_h. \quad \text{In } h, \text{ we have: } X_h = X_g, \text{ the domain of } g.$$
$$\underset{h}{\overbrace{\qquad\qquad}}$$

So the domain of h is: $x \geq 4$. And next, checking to see if the range of g is a subset of the domain of f, we get: $x \geq 4 \Rightarrow x - 4 \geq 0 \Rightarrow \sqrt{x-4} \geq 0 \Rightarrow y \geq 0$, which is the range of g, and equals the domain of f. So h can be defined in the domain of g.

Let's see though, what will happen in this case.

We have: $y = g(x) = \sqrt{x - 4}, y = f(x) = x^2$, and $y = h(x) = (f \bullet g)(x) = f(g)$.

So we get: $h(x) = f(g) = g^2 = x - 4$, which has no problem with the domain X_g: $x \geq 4$.

Thus, we get: $y = h(x) = x - 4$ for $x \geq 4$. Let's next, check the ranges of f and h.

Beginning with the range of h, we get: $x \geq 4 \Rightarrow x - 4 \geq 0 \Rightarrow y \geq 0$.

And next, getting the range of f, we get: $x \geq 0 \Rightarrow x^2 \geq 0 \Rightarrow y \geq 0$.

So h can be called a function from the domain of g to the range of f.

Now, let's for next instance, find $y = h(x) = (g \bullet f)(x)$ assuming $y = g(x) = x^2$ for $0 \leq x \leq 2$, and $y = f(x) = \sqrt{1 - x^2}$ for $0 \leq x \leq 1$.   To begin with, we have:

$$X_f \xrightarrow{\ f\ } Y_f \xrightarrow{\ g\ } Y_h \quad \Rightarrow \quad X_f \xrightarrow{\ g \bullet f\ } Y_h. \quad \text{In } h, \text{ we have: } X_h = X_f, \text{ of course.}$$
$$\underset{h}{\overbrace{\qquad\qquad}}$$

So the domain of h is: $0 \leq x \leq 1$.

And next, checking to see if the range of f is a subset of the domain of g, we get:

$$0 \le x \le 1 \Rightarrow 0 \le x^2 \le 1 \Rightarrow -1 \le -x^2 \le 0 \Rightarrow 0 \le 1 - x^2 \le 1 \Rightarrow 0 \le \sqrt{1 - x^2} \le 1 \Rightarrow 0 \le y \le 1,$$

which is the range of f. And the domain of g is: $0 \le x \le 2$.

So the range of f is a subset of the domain of g, and thus, h can be defined in the domain of f.

Let's see though, what will happen in this case.

We have: $y = f(x) = \sqrt{1 - x^2}$, $y = g(x) = x^2$, and $y = h(x) = (g \bullet f)(x) = g(f)$.

So we get: $h(x) = g(f) = 1 - x^2$, which has no problem with the domain X_f: $0 \le x \le 1$.

Let's next, check the ranges of g and h.

Beginning with the range of g, we get: $0 \le x \le 2 \Rightarrow 0 \le x^2 \le 4 \Rightarrow 0 \le y \le 4$.

And next, finding the range in h, we get:

$$0 \le x \le 1 \Rightarrow 0 \le x^2 \le 1 \Rightarrow -1 \le -x^2 \le 0 \Rightarrow 0 \le 1 - x^2 \le 1 \Rightarrow 0 \le y \le 1,$$

which is not the range in g. So the range in h is just a part of the range in g.

That's because the range of the function blends into is a part of the domain of the other function.

We know that the range of the function blends into is used as the domain of the other function.

The function blends into is f, where the range is: $0 \le y \le 1$, and the other function is g, where the domain is: $0 \le x \le 2$.

And therefore, h is not a composite function from the domain of the function blends into to the range of the other function.

However, h is a function working perfectly for the domain of the function blends into.

So we can still get the composite function that is perfectly defined in the domain of the function blends into.
And thus, in this case, too, we can get a composite function working nicely.

For one more instance, let's find $y = q(x) = (h \bullet (g \bullet f))(x)$ using the functions below:

$y = f(x) = x + 1$ for $1 \leq x \leq 2$, $y = g(x) = 2x + 1$ for $2 \leq x \leq 3$, and

$y = h(x) = 9 - x$ for $5 \leq x \leq 7$.

Without showing variables, we can put q briefly this way, too: $q = h \bullet g \bullet f$, which is thus, equal to $q = h \bullet (g \bullet f)$, and is not equal to $q = (h \bullet g) \bullet f$, of course. And q is a blend of f, g, and h, but is not just a blend but a mixture made in the sequential manner as below.

Now, finding $q = (h \bullet (g \bullet f))(x)$, we want to do $g \bullet f$, first.

So assuming $p = g \bullet f$, we want to first, combine f and g to get p, which is the blend of f and g, which is more specifically, a composite function where f blends into g.

And next, the function p blends into the other function h, then we get q, which is a composite function, where p blends into h. So to begin with, we have:

$$X_f \xrightarrow{\ f\ } Y_f \xrightarrow{\ g\ } Y_p \quad \Rightarrow \quad X_f \xrightarrow{\ g \bullet f\ } Y_p. \quad \text{And in } p, \text{ we have: } X_p = X_f, \text{ of course.}$$
$$\underbrace{\qquad\qquad}_{p}$$

So the domain of p is the domain of f, and therefore, is: $1 \leq x \leq 2$.

Now, we have: $y = f(x) = x + 1$, $y = g(x) = 2x + 1$, and $y = p(x) = (g \bullet f)(x) = g(f)$.

So we get: $p(x) = g(f) = 2f + 1 = 2(x + 1) + 1 = 2x + 3$, which has no problem with the domain where: $1 \leq x \leq 2$.

And next, we want to blend p into the function h. Prior to such blending though, we want to make sure that the range of p fits the domain of h, which is: $5 \leq x \leq 7$.

That is, the range of p needs to be equal to or a part of the domain of h.

So finding the range of p, we get:

$1 \leq x \leq 2 \Rightarrow 2 \leq 2x \leq 4 \Rightarrow 5 \leq 2x + 3 \leq 7 \Rightarrow 5 \leq y \leq 7$, which equals the domain of h.

So now we can move on to: $y = q(x) = (h \bullet p)(x)$, because $p = g \bullet f$.

Then, we have:

$$X_p \xrightarrow{\ p\ } Y_p \xrightarrow{\ h\ } Y_q \quad \Rightarrow \quad X_f \xrightarrow{\ h \bullet p\ } Y_h.$$

with the curved arrow labeled q from X_p to Y_q. And we have: $X_q = X_p = X_f$, of course.

So the domain of q is: $1 \leq x \leq 2$.

Now, we have: $y = p(x) = 2x + 3$, $y = h(x) = 9 - x$, and $y = q(x) = (h \bullet p)(x) = h(p)$.

So we get: $q(x) = h(p) = 9 - p = 9 - (2x + 3) = -2x + 6$, which has no problem with the domain where: $1 \leq x \leq 2$.

Therefore, we get: $y = q(x) = -2x + 6$ for $1 \leq x \leq 2$.

And next, let's check the ranges in q and h.

Then, beginning with the range of h, we get:

$1 \leq x \leq 2 \Rightarrow 2 \leq 2x \leq 4 \Rightarrow -4 \leq -2x \leq -2 \Rightarrow 2 \leq -2x + 6 \leq 4 \Rightarrow 2 \leq y \leq 4$.

And next, the domain of h is: $5 \leq x \leq 7$, so finding the range of h, we get:

$5 \leq x \leq 7 \Rightarrow -7 \leq -x \leq -5 \Rightarrow 2 \leq 9 - x \leq 4 \Rightarrow 2 \leq y \leq 4$, which equals the range of q.

So the function q is: $y = q(x) = -2x + 6$ for $1 \leq x \leq 2$, which is a composite function from the domain of f to the range of h.

And therfore, the composite function q has all the inputs of f, and produces all the outputs of h.

So q is a blend of three functions f, g, and h.

And the blending sequence matters.

So more specifically, q is a composite function, where f blends into g, then the blend $(g \bullet f)$ blends into the other function h.

Let's now put in a graph all the functions f, g, h, p, and q.

$$y = f(x) = x + 1 \text{ for } 1 \leq x \leq 2$$
$$y = g(x) = 2x + 1 \text{ for } 2 \leq x \leq 3$$
$$y = p(x) = (g \bullet f)(x) = 2x + 3 \text{ for } 1 \leq x \leq 2$$
$$y = h(x) = 9 - x \text{ for } 5 \leq x \leq 7$$
$$y = q(x) = (h \bullet (g \bullet f))(x) = (h \bullet p)(x) = -2x + 6 \text{ for } 1 \leq x \leq 2$$

Fig. 5.3

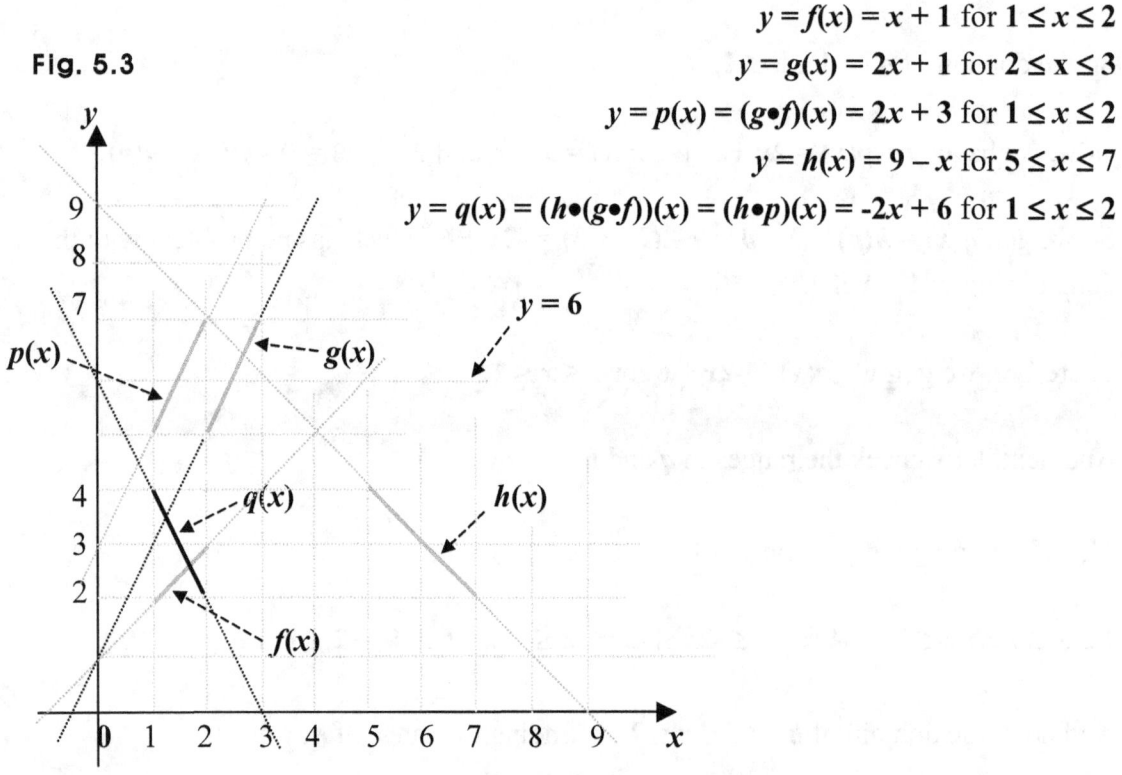

Examples 1 in Composite Functions

Find each of all the composite functions below using the definitions as follows:
$f: x \longrightarrow x$, $g: x \longrightarrow 2x^2 + 1$, and $h: x \longrightarrow 3x^3 + 2$.

0. $g \bullet f$

1. $f \bullet g$

2. $g \bullet h$

3. $h \bullet g$

4. $h \bullet (g \bullet f)$

5. $(h \bullet g) \bullet f$

6. $h \bullet (g \bullet g)$

7. $(h \bullet g) \bullet g$

Suggestions or Solutions
To the **Problem** in the Example **0**

Find $g \bullet f$ using the definition as follows: $f: x \longrightarrow x$, and $g: x \longrightarrow 2x^2 + 1$.

$f(x) = x$, and $g(x) = 2x^2 + 1$, so $g \bullet f = g\{f(x)\} = 2\{f(x)\}^2 + 1 = 2x^2 + 1$ for x real.

If not quite sure of the idea behind the processes above, follow the steps below:

To begin with, if not specified, the domain is assumed to be a set of all real numbers, or the set of all numbers the function can be defined for. More specifically, it is the set of all numbers the expression can hold for. For instance, $\sqrt{x+1}$ can hold for $x = 1$, but cannot hold for $x = -2$, because what's inside the square root cannot be negative.

Next, if R is a set of all real numbers, the expressions of f and g both can hold for R. So the domain of f is R, so is the range, and so is the domain of g.

And thus, the function f is defined in R, since R is the domain.

Where then, is a composite function defined?

A composite function is defined in the domain of the function blends into. Also, the range of the function blends into has to be a subset of the domain of the other function.

Now, if we set: $p = g \bullet f$, p is a composite function, where f is the function blends into.

So p has to be defined in the domain of f, and the range of f has to be a subset of the domain of g. And for instance, a subset of a set A is equal to or a part of the set A.

We know the range of f is R, and so is the domain of g. That is, the range of f is a subset of the domain of g. Thus, we should be able to get the composite function p in this case.

So assuming $p = g \bullet f$, we have: $y = p(x) = (g \bullet f)(x)$, and can put it this way, too:

$$X_f \xrightarrow{\ f\ } Y_f \xrightarrow{\ g\ } Y_p \quad \Rightarrow \quad X_f \xrightarrow{\ g \bullet f\ } Y_p. \quad \text{And of course, } X_f = X_p, \text{ and } p\text{'s range is } Y_p.$$

$$\underset{p}{\overbrace{}}$$

Now, in the diagram above, X_f is the domain of f, the function blends into, and thus, is the domain of p. So lets now get the expression of p.

We have: $y = p(x) = (g \bullet f)(x) = g(f)$, $y = g(x) = 2x^2 + 1$, and $y = f(x) = x$.

So we get: $p(x) = g(f) = 2f^2 + 1 = 2x^2 + 1$, since $f(x) = x$.

Thus, we get: $y = (g \bullet f)(x) = 2x^2 + 1$ for x real, which is the same as the function g given.

We know f is an identity function. So getting a composite function mixing two functions, we just get the other function if the function blends into is an identity function. It doesn't matter though, which one is the function blends into.

That is, if one of the two is an identity function, we just get the other function.

In short:

$f(x) = x$, and $g(x) = 2x^2 + 1$, so $g \bullet f = g\{f(x)\} = 2\{f(x)\}^2 + 1 = 2x^2 + 1$ for x real.

Suggestions or Solutions
To the Problem in the Example 1

Find $f \bullet g$ using the definition as follows: $f: x \longrightarrow x$, and $g: x \longrightarrow 2x^2 + 1$.

$f(x) = x$, and $g(x) = 2x^2 + 1$, so $f \bullet g = f\{g(x)\} = g(x) = 2x^2 + 1$ for x real.

If not quite sure of the idea behind the processes above, follow the steps below:

In this problem, too, we want to get a composite function mixing two functions, which are though, the same ones we used in the previous problem. In this problem though, the mixing sequence is the opposite.

So first, assuming $y = p(x) = (f \bullet g)(x)$, we have:

$$X_g \xrightarrow{\quad g \quad} Y_g \xrightarrow{\quad f \quad} Y_p \quad \Rightarrow \quad X_g \xrightarrow{\quad f \bullet g \quad} Y_p. \quad \text{And in } p \text{, we have: } X_p = X_g, \text{ of course.}$$
$$\xrightarrow{\qquad p \qquad}$$

So this time, of the two functions f and g, g is the one blends into. And thus, the range of g needs to fit the domain of f.

In the previous problem, we've already found that the range of g is: $y \geq 1$, and that the domain of f is R, which is a set of all real numbers. So the range of the functions blends into is a subset of the domain of the other function, and thus, we can get the composite function that can be defined in the domain of g, which is R, of course.

So next, we want to get the expression of p.

We have: $y = p(x) = (f \bullet g)(x) = f(g)$, $y = g(x) = 2x^2 + 1$, and $y = f(x) = x$.

Thus, we get: $p(x) = f(g) = g$, since $f(x) = x$. And we know: $g = 2x^2 + 1$.

So we get: $p(x) = 2x^2 + 1$, which has, of course, no problem with the domain R, and therefore, we get: $y = (f \bullet g)(x) = 2x^2 + 1$ for x real, which is again, the same as the function g given. And we know f is an identity function.

So considering the previous problem, we can be sure that getting a composite function using two functions, one of which is an identity function, we just get the other function. That is, if f is an identity function, we just get: $f \bullet g = g \bullet f = g$.

Suggestions or Solutions
To the **Problem** in the Example **2**

Find $g \bullet h$ **using the definition as follows:** $g: x \longrightarrow 2x^2 + 1$, and $h: x \longrightarrow 3x^3 + 2$.

$g(x) = 2x^2 + 1$, and $h(x) = 3x^3 + 2$, so $g \bullet h = g\{h(x)\} = g(3x^3 + 2) = 2(3x^3 + 2)^2 + 1$ for x real.

If not quite sure of the idea behind the processes above, follow the steps below:

To begin with, assuming $y = p(x) = (g \bullet h)(x)$, we have:

$$X_h \xrightarrow{\ h\ } Y_h \xrightarrow{\ g\ } Y_p \quad \Rightarrow \quad X_h \xrightarrow{\ g \bullet h\ } Y_p. \quad \text{And in } p, \text{ we have: } X_p = X_h, \text{ of course.}$$

So of the two functions h and g we want to mix, h is the function blends into. And thus, the range of h has to fit the domain of g.

The range of h is: $y \geq 2$, and the domain of g is R, which is a set of all real numbers.

So the range is a subset of the domain, and thus, we can get the composite function that can be defined in the domain of the function blends into.

And the domain is X_h, which is R, which is the domain of h, of course.

Thus next, we want to get the expression of p.

Now, we have: $y = p(x) = (g \bullet h)(x) = g(h)$, $y = g(x) = 2x^2 + 1$, and $y = h(x) = 3x^3 + 2$.

So we get: $p(x) = g(h) = 2h^2 + 1$, since $g(x) = 2x^2 + 1$. And thus, $p(x) = 2(3x^3 + 2)^2 + 1$, which is perfectly defined for $x \in R$, which is the domain of the function h blends into.

And next, finding the ranges of p and g, we can see the range of g is: $y \geq 1$, so is the range of p. So p is a composite function from the domain of h to the range of g.

How come though, the range of p is: $y \geq 1$?

The expression in p is just an algebraic expression of degree 6, and we don't have to expand (or simplify) the expression if not necessary, of course.

Expanding it though, we get: $2(3x^3 + 2)^2 + 1 = 2(9x^6 + 12x^3 + 4) + 1 = 18x^6 + 24x^3 + 9$.

Now, in the expression of p, we have: $2(3x^3 + 2)^2$, which is however, ≥ 0.

That's because:

First, x^3 can be any number negative as well as 0 or positive, since $x \in R$, and thus, can be -2. So $3x^3 + 2$ can be 0.

Next, $(3x^3 + 2)^2$ cannot be < 0, because any number squared is ≥ 0.

And thus, we get: $2(3x^3 + 2)^2 + 1 \geq 1 \Rightarrow y \geq 1$, which is the range of p.

And of course, setting $t = x^3$, we just get: $2(3x^3 + 2)^2 + 1 = 2(3t + 2)^2 + 1 \geq 1$, too.

Therefore, we get: $y = (g \bullet h)(x) = 2(3x^3 + 2)^2 + 1$ for x real.

In short:

$g(x) = 2x^2 + 1$, and $h(x) = 3x^3 + 2$, so $g \bullet h = g\{h(x)\} = g(3x^3 + 2) = 2(3x^3 + 2)^2 + 1$ for x real.

Suggestions or Solutions
To the **Problem** in the Example **3**

Find $h \bullet g$ using the definition as follows: $g: x \longrightarrow 2x^2 + 1$, and $h: x \longrightarrow 3x^3 + 2$.

$g(x) = 2x^2 + 1$, and $h(x) = 3x^3 + 2$, so $h \bullet g = h\{g(x)\} = h(2x^2 + 1) = 3(2x^2 + 1)^3 + 2$ for x real.

If not quite sure of the idea behind the processes above, follow the steps below:

To begin with, assuming $y = p(x) = (h \bullet g)(x)$, we have:

$$X_g \xrightarrow{\ g\ } Y_g \xrightarrow{\ h\ } Y_p \quad \Rightarrow \quad X_g \xrightarrow{\ h \bullet g\ } Y_p. \quad \text{And in } p, \text{ we have: } X_p = X_g, \text{ of course.}$$

So on the contrary to the previous problem, of the two functions h and g we want to mix, which one is the function blends into?

It is the function g.

And thus, the range of g has to fit the domain of h.

What then, is the range of g?

The range of g is: $y \geq 1$, and the domain of h is R, which is a set of all real numbers.

So the range of g is a subset of the domain of h, and thus, we can get the composite function p that can be defined in the domain of the function blends into.

And the domain of p is X_g, which is R, which is of course, the domain of g.

So next, we want to get the expression of p.

We have: $y = p(x) = (h \bullet g)(x) = h(g)$, $y = g(x) = 2x^2 + 1$, and $y = h(x) = 3x^3 + 2$.

So we get: $p(x) = h(g) = 3g^3 + 2$, since $h(x) = 3x^3 + 2$. And we have: $g = 2x^2 + 1$.

Thus, we get: $p(x) = 3(2x^2 + 1)^3 + 2$, which can be defined for $x \in R$, which is X_g, which is the function blends into. And let's next, check the ranges of p and g.

Beginning with the range of g, we get: $y \geq 1$. And next, finding the range of p, we get:

$2x^2 + 1 \geq 1 \Rightarrow (2x^2 + 1)^3 \geq 1 \Rightarrow 3(2x^2 + 1)^3 \geq 3 \Rightarrow 3(2x^2 + 1)^3 + 2 \geq 5 \Rightarrow y \geq 5$.

So the range of p is just a part of the range of g. That's simply because the range in the function blends into is a part of the domain of the other function.

And thus, p is not a function from the domain of the function blends into to the range of the other function.

The function p is however, a composite function perfectly working in the domain of the function blends into.

Therefore, the solution to this problem is: $y = (h \bullet g)(x) = 3(2x^2 + 1)^3 + 2$ for x real.

And comparing the expression above to the one in the previous problem, we can see again, that it's *not* the case where we get: $h \bullet g = g \bullet h$ if g and h are the two functions given.

So we can say that in general, we get: $h \bullet g \neq g \bullet h$.

In short:

$g(x) = 2x^2 + 1$, and $h(x) = 3x^3 + 2$, so $h \bullet g = h\{g(x)\} = h(2x^2 + 1) = 3(2x^2 + 1)^3 + 2$ for x real.

Suggestions or Solutions
To the **Problem** in the Example **4**

Find $h \bullet (g \bullet f)$ using the definition below:
$f: x \longrightarrow x$, $g: x \longrightarrow 2x^2 + 1$, and $h: x \longrightarrow 3x^3 + 2$.

$f(x) = x$, $g(x) = 2x^2 + 1$, and $h(x) = 3x^3 + 2$.

So first, we get: $g \bullet f = g\{f(x)\} = 2f^2 + 1 = 2x^2 + 1$ for x real.

Thus, $h \bullet (g \bullet f) = h \bullet g = h(g) = h(2x^2 + 1) = 3(2x^2 + 1)^3 + 2 = 24x^6 + 36x^4 + 18x^2 + 5$ for x real.

If not quite sure of the idea behind the processes above, follow the steps below:

We can get a composite function mixing more than two functions. We don't however, just mix together all at once. We mix two at a time sequential manner.

Assuming for instance, mixing three functions, we mix first, two of the three, and get the blend of the two.

Then, mixing together the blend and the other function, we get a composite function, which is therefore, the blend of all the three functions.

We cannot do so however, if all the domains and ranges facing each other do not fit each other. That is, of the two functions to mix, if the range of the function blends into is not a subset of the domain of the other function, the two functions don't mix, so we get no composite function.

To begin with, $g \bullet f$ is a composite function, which is thus, a function, too, so calling it p, we get: $h \bullet (g \bullet f) = h \bullet k$, which is a composite function, too, and is a function, also, so calling it q, we get: $q = h \bullet k$. And thus, we have: $q = h \bullet (g \bullet f)$.

And the same is true, too, for $h \bullet g \bullet f$, which is therefore, equal to $h \bullet (g \bullet f)$.

So making a composite function using more than two functions, we combine two at a time in sequential manner as stated above. Then, the last composite function is the one we want, and is a blend of all the functions used. And thus, we want to begin with $g{\bullet}f$.

So first, assuming $y = p(x) = (g{\bullet}f)(x)$, we have:

$$X_f \xrightarrow{\ f\ } Y_f \xrightarrow{\ g\ } Y_p \quad \Rightarrow \quad X_f \xrightarrow{\ g{\bullet}f\ } Y_p \quad \text{In } p, \text{ we have: } X_f = X_p, \text{ of course.}$$

And assuming next, R is a set of all real numbers, we need to assume that if the domain is unspecified in a function, it is a set of all real numbers.

And if in a function, the expression is a monomial or polynomial as the one in f, g or h, the domain is R. It is *not* always the case however, the range is R, too. The range of h is R, but that of g is: $y \geq 1$.

And a composite function is defined in the domain of the function blends into. Also, the range of the function blends into has to be a subset of the domain of the other function.

Now, the range of f is R, so is the domain of g, so we can get the composite function p. And f is the function blends into, and its domain is R, so the domain of p is R.

Thus next, we want to get the expression of p.

We have: $y = p(x) = (g{\bullet}f)(x) = g(f)$, $y = g(x) = 2x^2 + 1$, and $y = f(x) = x$, which is an identity function. So we just get: $p(x) = g(x) = 2x^2 + 1$ for x real.

So next, we want to move on to $h{\bullet}(g{\bullet}f)$, which is equal to $h{\bullet}p$.

Thus first, assuming $y = q(x) = (h{\bullet}p)(x)$, we have:

$$X_p \xrightarrow{\ p\ } Y_p \xrightarrow{\ h\ } Y_q \quad \Rightarrow \quad X_p \xrightarrow{\ h{\bullet}p\ } Y_q \quad \text{In } q, \text{ we have: } X_q = X_p = X_f, \text{ of course.}$$

So again, a composite function is defined in the domain of the function blends into. And also, the range of the function blends into has to be a subset of the domain of the other function. We want to check this every time we do blending.

Thus first, p is the function blends into, so q has to be composite function defined in the domain of p, where the domain is originally the domain of f.

And next, the range of p is: $y \geq 1$, and the domain of h is R. So we can get the composite function q, since the range of p is a subset of the domain of h.

So next, we want to get the expression of q.

We have: $y = q(x) = (h \bullet p)(x) = h(p)$, $y = p(x) = 2x^2 + 1$, and $y = h(x) = 3x^3 + 2$.

So we get: $q(x) = h(p) = 3p^3 + 2$, since $h(x) = 3x^3 + 2$. And we have: $p = 2x^2 + 1$.

Thus, we get: $q(x) = 3(2x^2 + 1)^3 + 2$. And we know $q = h \bullet (g \bullet f)$, and the domain of q is R.

So we get: $y = h \bullet (g \bullet f)(x) = 3(2x^2 + 1)^3 + 2$ for x real.

In the problem 3, we've found that if $x \in R$, we get: $3(2x^2 + 1)^3 + 2 \geq 5$.

So the range of q is: $y \geq 5$, which is thus, Y_q in the diagram above.

However, the domain of h is: R. So q is not a function from the domain of f to the range of g. That's because the range of p, the function blends into is not equal to the domain of the other function h.

However, q is a composite function perfectly working in the domain of the function blends into. So the solution to this problem is q.

In short:

$f(x) = x$, $g(x) = 2x^2 + 1$, and $h(x) = 3x^3 + 2$.

So first, we get: $g \bullet f = g\{f(x)\} = 2f^2 + 1 = 2x^2 + 1$.

Thus, $h \bullet (g \bullet f) = h \bullet g = h(g) = h(2x^2 + 1) = 3(2x^2 + 1)^3 + 2 = 24x^6 + 36x^4 + 18x^2 + 5$.

Suggestions or Solutions
To the Problem in the Example 5

Find $(h \bullet g) \bullet f$ using the definition below:

$f\colon x \longrightarrow x$, $g\colon x \longrightarrow 2x^2 + 1$, and $h\colon x \longrightarrow 3x^3 + 2$.

$f(x) = x$, $g(x) = 2x^2 + 1$, and $h(x) = 3x^3 + 2$.

Thus first, we get: $(h \bullet g) \bullet f = h \bullet g$, because f is an identity function.

So $(h \bullet g) \bullet f = h \bullet g = h(g) = 3g^3 + 2 = 3(2x^2 + 1)^3 + 2 = 24x^6 + 36x^4 + 18x^2 + 5$ for x real.

If not quite sure of the idea behind the processes above, follow the steps below:

To begin with, in $(h \bullet g) \bullet f$, we know $h \bullet g$ is a composite function.

So assuming $p = h \bullet g$, we get: $(h \bullet g) \bullet f = p \bullet f$.

And calling it q, we get: $q = p \bullet f$, which is of course, equal to $(h \bullet g) \bullet f$.

So we have: $q = (h \bullet g) \bullet f$.

And thus, we want to begin with $h \bullet g$, where g is the function mixes into h.

So first, assuming $y = p(x) = (h \bullet g)(x)$, we have:

$$X_g \xrightarrow{\;g\;} Y_g \xrightarrow{\;h\;} Y_p \quad \Rightarrow \quad X_g \xrightarrow{\;h \bullet g\;} Y_p \quad \text{In } p \text{, we have: } X_p = X_g \text{, of course.}$$

with the arrow labeled p underneath from X_g to Y_p.

And a composite function is defined in the domain of the function blends into. Also, the range of the function blends into has to be a subset of the domain of the other function.

Now, the function g blends into the function h, the range of g is $y \geq 1$, and the domain of h is R, which is a set of all real numbers.

So we can get the composite function p, defined in the domain of g, and the domain is R.

Thus next, we want to get the expression of p.

How then, do we get the expression of p?

We have: $y = p(x) = (h \bullet g)(x) = h(g)$, $y = g(x) = 2x^2 + 1$, and $y = h(x) = 3x^3 + 2$.

Thus, we get: $p(x) = h(g) = 3g^3 + 2 = 3(2x^2 + 1)^3 + 2$ for x real.

So next, we want to move on to $(h \bullet g) \bullet f$, which is equal to $p \bullet f$.

Thus first, assuming $y = q(x) = (p \bullet f)(x)$, we have:

$$X_f \xrightarrow{\ f\ } Y_f \xrightarrow{\ p\ } Y_q \ \underset{q}{\overset{}{\Rightarrow}} \ X_f \xrightarrow{\ p \bullet f\ } Y_p \quad \text{In } q, \text{ we have: } X_q = X_f, \text{ of course.}$$

Now, the function p blends into the function f, the range of f is R, a set of all real numbers, and the domain of p is R, too. So we can get the composite function q, defined in the domain of f, and the domain is R. And next, we want to get the expression of q.

We have: $y = q(x) = (p \bullet f)(x) = p(f)$, $y = p(x) = 3(2x^2 + 1)^3 + 2$, and $y = f(x) = x$, which is an identity function.

So we just simply get: $y = q(x) = p(x) = 3(2x^2 + 1)^3 + 2$ for x real.

In short:

$f(x) = x$, $g(x) = 2x^2 + 1$, and $h(x) = 3x^3 + 2$.

Thus first, we get: $(h \bullet g) \bullet f = h \bullet g$, because f is an identity function.

So $(h \bullet g) \bullet f = h \bullet g = h(g) = 3g^3 + 2 = 3(2x^2 + 1)^3 + 2 = 24x^6 + 36x^4 + 18x^2 + 5$ for x real.

Suggestions or Solutions
To the **Problem** in the Example 6

Find $h \bullet (g \bullet g)$ using the definition as follows: $g: x \longrightarrow 2x^2 + 1$, and $h: x \longrightarrow 3x^3 + 2$.

$g(x) = 2x^2 + 1$, and $h(x) = 3x^3 + 2$.

So first, $g \bullet g = g\{g(x)\} = 2g^2 + 1 = 2(2x^2 + 1)^2 + 1$.

Next, setting $p = g \bullet g$, we get: $h \bullet (g \bullet g) = h \bullet p = h(p) = 3\{2(2x^2 + 1)^2 + 1\}^3 + 2$.

If not quite sure of the idea behind the processes above, follow the steps below:

Making a composite function, we can use one function only, too.
And we can use the same function as many times as needed. It is the case, of course, only if the range of each function blends into is the domain of the next function.

So $g \bullet g$ is a composite function, and calling it p, we get: $h \bullet (g \bullet g) = h \bullet p$.

And thus, to begin with, assuming $y = p(x) = (g \bullet g)(x)$, we have:

$$X_g \xrightarrow{\ g\ } Y_g \xrightarrow{\ g\ } Y_g \quad \Rightarrow \quad X_g \xrightarrow{\ g \bullet g\ } Y_p \quad \text{In } p, \text{ we have: } X_p = X_g, \text{ of course.}$$
$$\underset{p}{\underbrace{\qquad\qquad\qquad}}$$

Now, the function g blends into itself, the range of g is $y \geq 1$, and the domain of g is R, a set of all real numbers. So we can get the composite function p, defined in the domain of g, and the domain is R. And next, we want to get the expression of p. How?

We have: $y = p(x) = (g \bullet g)(x) = g(g)$, and $y = g(x) = 2x^2 + 1$.

So we get: $p(x) = g(g) = 2g^2 + 1 = 2(2x^2 + 1)^2 + 1$ for x real. And the range is $y \geq 3$.

So next, we want to move on to $h \bullet (g \bullet g)$, which is equal to $h \bullet p$.

Thus first, assuming $y = q(x) = (h \bullet p)(x)$, we have:

$$X_p \xrightarrow{\;p\;} Y_p \xrightarrow{\;h\;} Y_q \quad \Rightarrow \quad X_g \xrightarrow{\;h \bullet p\;} Y_q \quad \text{In } q \text{, we have: } X_q = X_p = X_g, \text{ of course.}$$
$$\underset{q}{\curvearrowright}$$

Now, the function p blends into the function h, the range of p is $y \geq 3$, and the domain of h is R. So we can get the composite function q, defined in the domain of p, which is originally the domain of g, which is R. And next, we want to get the expression of q.

We have: $y = q(x) = (h \bullet p)(x) = h(p)$, $p(x) = 2(2x^2 + 1)^2 + 1$, and $h(x) = 3x^3 + 2$.

So we get: $y = q(x) = h(p) = 3p^3 + 2 = 3\{2(2x^2 + 1)^2 + 1\}^3 + 2$ for x real.

And checking the range of q, we get: $2x^2 + 1 \geq 1 \Rightarrow (2x^2 + 1)^2 \geq 1 \Rightarrow 2(2x^2 + 1)^2 \geq 2$

$\Rightarrow 2(2x^2 + 1)^2 + 1 \geq 3 \Rightarrow \{2(2x^2 + 1)^2 + 1\}^3 \geq 3^3 \Rightarrow 3\{2(2x^2 + 1)^2 + 1\}^3 \geq 3^4$

$\Rightarrow 3\{2(2x^2 + 1)^2 + 1\}^3 + 2 \geq 3^4 + 2 \Rightarrow y \geq 83$.

In short:

$g(x) = 2x^2 + 1$, and $h(x) = 3x^3 + 2$.

So first, $g \bullet g = g\{g(x)\} = 2g^2 + 1 = 2(2x^2 + 1)^2 + 1$.

Next, setting $p = g \bullet g$, we get: $h \bullet (g \bullet g) = h \bullet p = h(p) = 3\{2(2x^2 + 1)^2 + 1\}^3 + 2$ for x real.

Suggestions or Solutions
To the Problem in the Example 7

Find $(h \bullet g) \bullet g$ using the definition as follows: $g: x \longrightarrow 2x^2 + 1$, and $h: x \longrightarrow 3x^3 + 2$.

$f(x) = x$, $g(x) = 2x^2 + 1$, and $h(x) = 3x^3 + 2$.

So first, we get: $h \bullet g = h\{g(x)\} = 3g^3 + 2 = 3(2x^2 + 1)^3 + 2$. And setting $p = h \bullet g$, we get:

$(h \bullet g) \bullet g = p \bullet g = p(g) = 3(2g^2 + 1)^3 + 2 = 3\{2(2x^2 + 1)^2 + 1\}^3 + 2$ for x real.

If not quite sure of the idea behind the processes above, follow the steps below:

Finding $(h \bullet g) \bullet g$, we want to find $h \bullet g$ first.

So to begin with, assuming $y = p(x) = (h \bullet g)(x)$, we have:

In p, we have: $X_p = X_g$, of course.

Now, the function g blends into the function h, the range of g is $y \geq 1$, and the domain of h is R, a set of all real numbers. So we can get the composite function p, defined in the domain of g, which is R. Thus next, we want to get the expression of p.   How?

We have: $y = p(x) = (h \bullet g)(x) = h(g)$, $y = g(x) = 2x^2 + 1$, and $y = h(x) = 3x^3 + 2$.

So we get: $p(x) = h(g) = 3g^3 + 2 = 3(2x^2 + 1)^3 + 2$, because $g = 2x^2 + 1$.

And thus, $p(x) = 3(2x^2 + 1)^3 + 2$ for $x \in R$.

So we can now move on to $(h \bullet g) \bullet g$, which is equal to $p \bullet g$.

Thus first, assuming $y = q(x) = (p \bullet g)(x)$, we have:

$$X_g \xrightarrow{\quad g \quad} Y_g \xrightarrow{\quad p \quad} Y_q \quad \Rightarrow \quad X_g \xrightarrow{\quad p \bullet g \quad} Y_q \quad \text{In } q, \text{ we have: } X_q = X_g, \text{ of course.}$$

$$\underset{q}{\underbrace{\qquad\qquad\qquad}}$$

Now, the function g blends into the function p, the range of g is $y \geq 1$, and the domain of p is R. So we can get the composite function q, defined in the domain of g, which is R.

Thus next, we want to get the expression of q.

We have: $y = q(x) = (p \bullet g)(x) = p(g)$, $p(x) = 3(2x^2 + 1)^3 + 2$, and $g(x) = 2x^2 + 1$.

So we get: $y = q(x) = p(g) = 3(2g^2 + 1)^3 + 2 = 3\{2(2x^2 + 1)^2 + 1\}^3 + 2$, which is OK with the domain R.

Therefore, $y = \{(h \bullet g) \bullet g\}(x) = 3\{2(2x^2 + 1)^2 + 1\}^3 + 2$ for x real.

In the problem 6 though, we found this: $\{h \bullet (g \bullet g)\}(x) = 3\{2(2x^2 + 1)^2 + 1\}^3 + 2$ for x real.

So do we have: $f \bullet g \bullet h = f \bullet (g \bullet h) = (f \bullet g) \bullet h$?

We will see if it is the case in **Examples 2**.

In short:

$f(x) = x$, $g(x) = 2x^2 + 1$, and $h(x) = 3x^3 + 2$.

So first, we get: $h \bullet g = h\{g(x)\} = 3g^3 + 2 = 3(2x^2 + 1)^3 + 2$. And setting $p = h \bullet g$, we get:

$(h \bullet g) \bullet g = p \bullet g = p(g) = 3(2g^2 + 1)^3 + 2 = 3\{2(2x^2 + 1)^2 + 1\}^3 + 2$.

Examples 2 in Composite Functions

Using three functions below, check to see if $(h \bullet g) \bullet f = h \bullet (g \bullet f)$.

$f: x \longrightarrow x + 1$, $g: x \longrightarrow x^2 - 1$, and $h: x \longrightarrow x^3 + 2$.

And we will get to see some more examples, which are going to be though, quite similar to the one above.

Suggestions or Solutions
To the Problem in the Example Given

Using three functions below, check to see if $(h \bullet g) \bullet f = h \bullet (g \bullet f)$.

$f: x \longrightarrow x + 1$, $g: x \longrightarrow x^2 - 1$, and $h: x \longrightarrow x^3 + 2$.

To begin with, assuming R is a set of all real numbers, we have:

$f(x) = x + 1$ for $x \in R$, $g(x) = x^2 - 1$ for $x \in R$, and $h(x) = x^3 + 2$ for $x \in R$.

So beginning with $h \bullet (g \bullet f)$, we get:

First, $g \bullet f = g(f) = f^2 - 1 = (x + 1)^2 - 1$.

And next, setting $p = g \bullet f$, we get: $p(x) = (x + 1)^2 - 1$.

So we get: $h \bullet (g \bullet f) = h \bullet p = h(p) = p^3 + 2 = \{(x + 1)^2 - 1\}^3 + 2$ for x real.

Next, moving on to $(h \bullet g) \bullet f$, we get:

First, $h \bullet g = h(g) = g^3 + 2 = (x^2 - 1)^3 + 2$.

And next, setting $r = h \bullet g$, we get: $r(x) = (x^2 - 1)^3 + 2$.

So we get: $(h \bullet g) \bullet f = r \bullet f = r(f) = (f^2 - 1)^3 + 2 = \{(x + 1)^2 - 1\}^3 + 2$ for x real.

Therefore, we get: $h \bullet (g \bullet f) = (h \bullet g) \bullet f$.

If not quite sure of the idea behind the processes above, follow the steps below:

What we want to do in this example is to check to see if composite-function-operations are associative.

To begin with, finding $(h \bullet g) \bullet f$, we want to find $h \bullet g$ first.

So first, assuming $y = p(x) = (h \bullet g)(x)$, we have:

$$X_g \xrightarrow{\ g\ } Y_g \xrightarrow{\ h\ } Y_p \underset{p}{\quad} \Rightarrow \quad X_g \xrightarrow{\ h \bullet g\ } Y_p \quad \text{And in } p, \text{ we have: } X_p = X_g.$$

Now, the function g blends into the function h, the range of g is: $y \geq -1$, and the domain of h is R, a set of all real numbers.

So we can get the composite function p, defined in the domain of g, which is R. Thus next, we want to get the expression of p.

We have: $y = p(x) = (h \bullet g)(x) = h(g)$, $y = g(x) = x^2 - 1$, and $y = h(x) = x^3 + 2$.

So we get: $p(x) = h(g) = g^3 + 2 = (x^2 - 1)^3 + 2$ for x real.

And we can now move on to $(h \bullet g) \bullet f$, which is equal to $p \bullet f$.

So first, assuming $y = q(x) = (p \bullet f)(x)$, we have:

$$X_f \xrightarrow{\ f\ } Y_f \xrightarrow{\ p\ } Y_q \underset{q}{\quad} \Rightarrow \quad X_f \xrightarrow{\ p \bullet f\ } Y_q \quad \text{And in } q, \text{ we have: } X_q = X_f.$$

Now, the function f blends into the function p, the range of f is R, and so is the domain of p. So we can get the composite function q, defined in the domain of f, which is R.

Thus next, we want to get the expression of q.

We have: $y = q(x) = (p \bullet f)(x) = p(f)$, $p(x) = (x^2 - 1)^3 + 2$, and $f(x) = x + 1$.

So we get: $y = q(x) = p(f) = (f^2 - 1)^3 + 2 = \{(x + 1)^2 - 1\}^3 + 2$, which is OK in R.

And thus, $y = \{(h \bullet g) \bullet f\}(x) = \{(x + 1)^2 - 1\}^3 + 2$ for x real.

And next, finding $h \bullet (g \bullet f)$, we want to find $g \bullet f$ first.

So first, assuming $y = r(x) = (g \bullet f)(x)$, we have:

$$X_f \xrightarrow{\ f\ } Y_f \xrightarrow{\ g\ } Y_r \quad \Rightarrow \quad X_f \xrightarrow{\ g \bullet f\ } Y_r \quad \text{And in } r, \text{ we have: } X_r = X_f.$$

Now, the function f blends into the function g, the range of f is R, and so is the domain of g. So we can get the composite function r, defined in R.

Thus next, we want to get the expression of r.

We have: $y = r(x) = (g \bullet f)(x) = g(f)$, $y = g(x) = x^2 + 1$, and $y = f(x) = x + 1$.

So we get: $r(x) = g(f) = f^2 + 1 = (x + 1)^2 + 1$ for x real.

And we can now move on to $h \bullet (g \bullet f)$, which is equal to $h \bullet r$.

So first, assuming $y = s(x) = (h \bullet r)(x)$, we have:

$$X_r \xrightarrow{\ r\ } Y_r \xrightarrow{\ h\ } Y_s \quad \Rightarrow \quad X_f \xrightarrow{\ h \bullet r\ } Y_s \quad \text{And in } s, \text{ we have: } X_s = X_r = X_f.$$

Now, the function r blends into the function h, the range of r is: $y \geq 1$, and the domain of h is R. So we can get the composite function s, defined in the domain R.

Thus next, we want to get the expression of s.

We have: $y = s(x) = (h \bullet p)(x) = h(p)$, $h(x) = x^3 + 2$, and $p(x) = (x + 1)^2 + 1$.

So we get: $y = s(x) = h(p) = p^3 + 2 = \{(x + 1)^2 + 1\}^3 + 2$, which is OK in R.

And thus, $y = \{h \bullet (g \bullet f)\}(x) = \{(x + 1)^2 + 1\}^3 + 2$ for x real.

Now, putting threads together, we have:

$y = \{(h \bullet g) \bullet f\}(x) = \{(x + 1)^2 + 1\}^3 + 2$ for x real.

$y = \{h \bullet (g \bullet f)\}(x) = \{(x + 1)^2 + 1\}^3 + 2$ for x real.

Therefore, we get: $h \bullet (g \bullet f) = (h \bullet g) \bullet f$.

In short:

To begin with, assuming R is a set of all real numbers, we have:

$f(x) = x + 1$ for $x \in R$, $g(x) = x^2 - 1$ for $x \in R$, and $h(x) = x^3 + 2$ for $x \in R$.

So beginning with $h \bullet (g \bullet f)$, we get:

First, $g \bullet f = g(f) = f^2 - 1 = (x + 1)^2 - 1$.

And next, setting $p = g \bullet f$, we get: $p(x) = (x + 1)^2 - 1$.

So we get: $h \bullet (g \bullet f) = h \bullet p = h(p) = p^3 + 2 = \{(x + 1)^2 - 1\}^3 + 2$ for x real.

Next, moving on to $(h \bullet g) \bullet f$, we get:

First, $h \bullet g = h(g) = g^3 + 2 = (x^2 - 1)^3 + 2$.

And next, setting $r = h \bullet g$, we get: $r(x) = (x^2 - 1)^3 + 2$.

So we get: $(h \bullet g) \bullet f = r \bullet f = r(f) = (f^2 - 1)^3 + 2 = \{(x + 1)^2 - 1\}^3 + 2$ for x real.

Therefore, we get: $h \bullet (g \bullet f) = (h \bullet g) \bullet f$.

So do we have: $h \bullet g \bullet f = h \bullet (g \bullet f) = (h \bullet g) \bullet f$?

(If you are really serious about composite functions, refer to the explanations below.)

Let's now check to see if it is the case using three functions below:

$f: A \longrightarrow B, g: C \longrightarrow D$, and $h: E \longrightarrow J$.

To begin with, finding $(h \bullet g) \bullet f$, we want to find $h \bullet g$ first.

So assuming $y = p(x) = (h \bullet g)(x)$, $D \subseteq E$, and this time, the range of p is K, we have:

$$C \xrightarrow{\;g\;} D \xrightarrow{\;h\;} K \quad \Rightarrow \quad C \xrightarrow{\;h \bullet g\;} K.$$

(with curved arrow p from C to K)

We assume $D \subseteq E$ so that the range of g, the function blends into, is a subset of the domain of h. Then, g can blend into h, and the domain of p is C. So we get:

$p: C \longrightarrow K$, that is, $h \bullet g: C \longrightarrow K$.

And next, we want to move on to $(h \bullet g) \bullet f$, which is equal to $p \bullet f$.

So assuming $y = q(x) = (p \bullet f)(x)$, $B \subseteq C$, and this time, the range of q is L, we have:

$$A \xrightarrow{\;f\;} C \xrightarrow{\;p\;} L \quad \Rightarrow \quad A \xrightarrow{\;p \bullet f\;} L.$$

(with curved arrow q from A to L)

We assume $B \subseteq C$ so that the range of f, the function blends into, is a subset of the domain of p. Then, f can blend into p, and the domain of q is A. So we get:

$q: A \longrightarrow L$, that is, $p \bullet f = (h \bullet g) \bullet f: A \longrightarrow L$.

And next, finding $h \bullet (g \bullet f)$, we want to find $g \bullet f$ first.

So assuming $y = r(x) = (g \bullet f)(x)$, $B \subseteq C$, and this time, the range of r is M, we have:

$$A \xrightarrow{\;f\;} A \xrightarrow{\;g\;} M \quad \Rightarrow \quad A \xrightarrow{\;g \bullet f\;} M.$$

(with curved arrow r from A to M)

We have already assume that $B \subseteq C$, so the range of f, the function blends into, is a subset of the domain of g. Thus, f can blend into g, and the domain of r is A. so we get:

$$r: A \longrightarrow M, \text{ that is, } g \bullet f: A \longrightarrow M.$$

And next, we want to move on to $h \bullet (g \bullet f)$, which is equal to $h \bullet r$.

So assuming $y = s(x) = (h \bullet r)(x)$, $M \subseteq E$, and this time, the range of s is N, we have:

$$A \xrightarrow{\ r\ } A \xrightarrow{\ h\ } N \quad \Rightarrow \quad A \xrightarrow{\ h \bullet r\ } N.$$

We assume $M \subseteq E$ so that the range of r, the function blends into, is a subset of the domain of h. Then, r can blend into h, and the domain of s is A. So we get:

$$s: A \longrightarrow N, \text{ that is, } h \bullet r = h \bullet (g \bullet f): A \longrightarrow N.$$

So putting threads together, we have: $(h \bullet g) \bullet f: A \longrightarrow L$, and $h \bullet (g \bullet f): A \longrightarrow N$

Then, it seems that we have ended up with $h \bullet (g \bullet f) \neq (h \bullet g) \bullet f$.

Let's take a look at though, what expression each of the two functions above has to get.

Then first, setting: $y = f(x)$, $y = g(x)$, and $y = h(x)$, we get:

$y = \{(h \bullet g) \bullet f\}(x) = (p \bullet f)(x) = p\{f(x)\}$, and $p(x) = (h \bullet g)(x) = h\{g(x)\}$.

So next, putting $f(x)$ into x in both of $h\{g(x)\}$ and $p(x)$, we get: $p\{f(x)\} = h[g\{f(x)\}]$.

Thus, we get: $(h \bullet g) \bullet f = p \bullet f = p\{f(x)\} = h[g\{f(x)\}]$.

More specifically, we can set: $\{(h \bullet g) \bullet f\}(x) = h[g\{f(x)\}]$.   In short, $(h \bullet g) \bullet f = h(g(f))$.

And making it more legible, we can put it this way, too, of course: $(h \bullet g) \bullet f = h\{g(f)\}$.

And next, moving on to $h \bullet (g \bullet f)$, we have: $h \bullet (g \bullet f) = h \bullet r = h(r)$, and $r = g \bullet f = g(f)$.

So we get: $h \bullet (g \bullet f) = h \bullet r = h(r) = h(g(f))$ since $r = g(f)$.

Thus, we get: $h \bullet (g \bullet f) = h(g(f))$.

And more legibly and specifically, we can set: $\{h \bullet (g \bullet f)\}(x) = h[g\{f(x)\}]$.

So we get: $\{(h \bullet g) \bullet f\}(x) = h[g\{f(x)\}]$, and $\{h \bullet (g \bullet f)\}(x) = h[g\{f(x)\}]$.

That is, the expressions in both functions are the same.

Now, putting threads together again, we have:

$(h \bullet g) \bullet f: A \longrightarrow L$, and $h \bullet (g \bullet f): A \longrightarrow N$, together with the fact that the expressions in both functions $(h \bullet g) \bullet f$ and $h \bullet (g \bullet f)$ are the same.

Then, we can see that we get: $L = N$.

(If taking a function as a system where an expression connects a domain to a range, we can say that if the domains are identical, and the expressions are identical, the functions are all the same, and thus, the ranges have to be identical, too.)

It is the case though, only if $D \subseteq E$, and $B \subseteq C$. We have:

$f: A \longrightarrow B, g: C \longrightarrow D$, and $h: E \longrightarrow J$.

So more specifically, we need to have: $D \subseteq E$, and $B \subseteq C$ so that $h \bullet (g \bullet f)$ and $(h \bullet g) \bullet f$ can exist, and then, we can get: $h \bullet (g \bullet f) = (h \bullet g) \bullet f$.

In other words, if $D \subseteq E$, and $B \subseteq C$, we have: $h \bullet g \bullet f = h \bullet (g \bullet f) = (h \bullet g) \bullet f$.

That is to say that composite-function-operations are associative if the domains and the ranges can work together.

What if the three functions as follows? $f: A \longrightarrow B, g: B \longrightarrow C$, and $h: C \longrightarrow D$.

Then, can we get: $h \bullet g \bullet f = h \bullet (g \bullet f) = (h \bullet g) \bullet f$ with no additional assumption on domains and ranges?

To begin with, finding $(h \bullet g) \bullet f$, we want to find $h \bullet g$ first.

Lets assume this time, the range of $h \bullet g$ is D, which is the range of h.

Then, assuming $y = p(x) = (h \bullet g)(x)$, we have:

$$B \xrightarrow{\ g\ } C \xrightarrow{\ h\ } D \quad \Rightarrow \quad B \xrightarrow{\ h \bullet g\ } D.$$
$$\text{(}p\text{)}$$

Assuming thus, the range of p is D, we get:

$$p: B \longrightarrow D, \text{ that is, } h \bullet g: B \longrightarrow D.$$

And next, we want to move on to $(h \bullet g) \bullet f$, which is equal to $p \bullet f$.

Let's assume again, the range of $p \bullet f$ is D, too, which is the range of h, of course.

Then, assuming $y = q(x) = (p \bullet f)(x)$, we have:

$$A \xrightarrow{\ f\ } B \xrightarrow{\ p\ } D \quad \Rightarrow \quad A \xrightarrow{\ p \bullet f\ } D.$$
$$\text{(}q\text{)}$$

Assuming thus, the range of q is D, we get: $q: A \longrightarrow D$, that is, $(h \bullet g) \bullet f: A \longrightarrow D$.

And next, finding $h \bullet (g \bullet f)$, we want to find $g \bullet f$ first.

This time, lets assume that the range of $g \bullet f$ is C, which is the range of g.

Then, assuming $y = r(x) = (g \bullet f)(x)$, we have:

$$A \xrightarrow{\ f\ } B \xrightarrow{\ g\ } C \quad \Rightarrow \quad A \xrightarrow{\ g \bullet f\ } C.$$
$$\text{(}r\text{)}$$

Assuming therefore, the range of *r* is *C*, we get:

$$r: A \longrightarrow C, \text{ that is, } g{\bullet}f: A \longrightarrow C.$$

And next, we want to move on to $h{\bullet}(g{\bullet}f)$, which is equal to $h{\bullet}r$.

Let's assume again though, the range of $h{\bullet}r$ is *D*, which is the range of *h*, of course.

Then, assuming $y = s(x) = (h{\bullet}r)(x)$, we have:

So assuming the range of *s* is *D*, we get: $q: A \longrightarrow D$, that is, $h{\bullet}(g{\bullet}f): A \longrightarrow D$.

And thus, putting threads together, we get: $h{\bullet}g{\bullet}f = h{\bullet}(g{\bullet}f) = (h{\bullet}g){\bullet}f$.

It is the case however, only if we make all the assumptions below.

The range of $h{\bullet}g$ is *D*, which is the range of *h*.

If $p = h{\bullet}g$, the range of $p{\bullet}f$ is *D*, too, which is the range of *h*.

The range of $g{\bullet}f$ is *C*, which is the range of *g*.

And if $r = g{\bullet}f$, the range of $h{\bullet}r$ is *D*, too, which is the range of *h*.

Examples 3 in Composite Functions

0.   Assuming $f(\frac{2x+1}{3}) = 2x+5$, find $f(7)$, and specify $f(\frac{3x-2}{2})$.

1.   Assuming $f(\frac{2x-1}{x}) = (1-2x^2)(3x^2+x+2)$, find $f(8)$.

2.   Assuming $f(x) = \frac{2x}{3x+1}$, and $g(x) = \frac{x}{3-2x}$, find the x-value that satisfies $(f \bullet g)(x) = f(x)$.

Suggestions or Solutions

To the **Problem** in the Example **0**

Assuming $f(\frac{2x+1}{3}) = 2x+5$, find $f(7)$, and specify $f(\frac{3x-2}{2})$.

Assuming first, $y = f(u)$, and $y = g(x) = \frac{2x+1}{3}$, we get: $y = f\{g(x)\} = f(\frac{2x+1}{3}) = 2x + 5$.

So setting: $u = \frac{2x+1}{3}$, we can get:

$u = \frac{2x+1}{3} \Rightarrow 2x+1 = 3u \Rightarrow x = \frac{3u-1}{2} \Rightarrow 2x+5 = 2 \cdot \frac{3u-1}{2} + 5 = 3u - 1 + 5 = 3u + 4.$

So we get: $f(u) = 3u + 4$, and thus, we get: $f(7) = 3 \cdot 7 + 4 = 25.$

And assuming next, $h(x) = \frac{3x-2}{2}$, we get: $f(\frac{3x-2}{2}) = f\{h(x)\}.$

And thus, putting $\frac{3x-2}{2}$ into u in $3u + 4$, we get: $3 \cdot \frac{3(3x-2)}{2} + 4 = \frac{9x-6+8}{2} = \frac{9x+2}{2}.$

So we get: $f(\frac{3x-2}{2}) = \frac{9x+2}{2}$ for x real.

If not quite sure of the idea behind the processes above, follow the steps below:

To begin with, what do we mean by $f(7)$?

It's the value of the output for 7 in case of the function f, and 7 is the input, of course.

So knowing the expression of f, we can readily find $f(7)$ putting 7 into the input variable in f, then evaluating the expression of f.

We are given though, this instead: $y = f(\frac{2x+1}{3}) = 2x + 5$ for x real. So what do we have?

The function given is a composite function. How come?

Getting a composite function, we blend a function into another function.

So assuming $g(x) = \frac{2x+1}{3}$, and blending g into f, we get: $y = f(\frac{2x+1}{3}) = 2x+5$. Not sure?

Assuming $y = f(u)$, and blending g into f, we get: $y = f(g)$. And we can give a name to it.

For instance, we can set: $k(x) = (f \bullet g)(x) = f(g)$.

Then, after putting the expression of g into every input variable in the expression of f, we get the expression of k, which is: $2x + 5$.

In short, putting the expression of g into u in $f(u)$, we get: $f(\frac{2x+1}{3}) = 2x+5$.

And of curse, the expression of f is an expression in terms of u.

So after putting the expression of g into every u in the expression of f, then simplifying the expression of f, we get: $2x + 5$, which is the expression of k.

And thus, we can set: $k(x) = (f \bullet g)(x) = f\{g(x)\} = 2x + 5$, where $g(x) = \frac{2x+1}{3}$.

How come though, u is the input variable of f?

In $f \bullet g$, f is not the function blends into, so we can use any letter as the input variable in f.

So putting the expression of g into the input variable in f, we get the expression of k.

And we know the expression of k is: $2x + 5$, which is therefore, certainly not the expression of f.

So we *cannot* just put 7 into x in $2x + 5$ to get $f(7)$.

What then, is the input variable in f? Or rather, what is the expression of f?

We have set: $y = f(u)$, where the expression is an expression in terms of u.

So putting the expression of g into every u in the expression of f, we get the expression of $f \bullet g$.

That is, putting $\frac{2x+1}{3}$ into u in $f(u)$, we get: $f\left(\frac{2x+1}{3}\right) = 2x+5$.   So what?

We can set: $u = \frac{2x+1}{3}$.

And we have this, too: $y = f\left(\frac{2x+1}{3}\right) = 2x + 5$.   So what do we get?

We get this, of course: $y = f(u) = 2x + 5$, in which however, the expression is not an expression in terms of u.   So what do we do with the equation, $u = \frac{2x+1}{3}$?

Solving it, we can put $2x + 5$, in terms of u.   How?

We have: $u = \frac{2x+1}{3}$, too. And x in $\frac{2x+1}{3}$ is the same as the very x in $2x + 5$.

So solving for x the equation $u = \frac{2x+1}{3}$, we get an expression in terms of u, which is of course, the solution to the equation.

And then, putting the expression into x in $2x + 5$, we get the expression of f.

That is, we get:

$$u = \frac{2x+1}{3} \Rightarrow 2x+1 = 3u \Rightarrow x = \frac{3u-1}{2} \Rightarrow 2x+5 = 2 \cdot \frac{3u-1}{2} + 5 = 3u - 1 + 5 = 3u + 4.$$

And thus, we get: $f(u) = 3u + 4$.

And of course, putting $\frac{2x+1}{3}$ back into u in $3u + 4$, we get this back: $2x + 5$.

That is, blending the expression of g into the expression of f, we get the expression of k, which is $f \bullet g$.

So now, putting 7 into u, we can simply get: $f(7) = 3 \cdot 7 + 4 = 25$.

Now, when blending g into f, we put g into u in $f(u)$, then we get the composite function $f \bullet g = f(g) = 3g + 4$, that is, putting $\frac{2x+1}{3}$ into u in $f(u)$, we can get the composite function: $(f \bullet g)(x) = 2x + 5$ for x real.

And we know in the composite function $f \bullet g$, g is the function blends into.

Where then, 7 in $f(7)$ is from?

The 7 in $f(7)$ is an output in the function g.

Then in g, what is the input, for which the output is 7?
In other words, for what value of x in $g(x)$, the output is 7?

We know: $u = \frac{2x+1}{3}$. So we get: $7 = \frac{2x+1}{3} \Rightarrow 2x = 20 \Rightarrow x = 10$.

And thus, we should get: $g(10) = 7$.

Checking to see if it really is the case, we simply get: $g(x) = \frac{2x+1}{3} \Rightarrow g(10) = \frac{2 \cdot 10 + 1}{3} = 7$.

So in the blending process, through the expressions in g and f, data flow the way below:

$10 \dashrightarrow x$ in $g(x) \dashrightarrow g(10) = 7 \dashrightarrow t$ in $f(t) \dashrightarrow f(7) = 25$.

- Now, let's next, move on to $f(\frac{3x-2}{2})$. What do we mean by $f(\frac{3x-2}{2})$, though?

It is another composite function, and in the function blends into, the expression is: $\frac{3x-2}{2}$.

And we have found that: $y = f(u) = 3u + 4$.

So if $h(x) = \frac{3x-2}{2}$, h is the function blends into, and thus, putting the expression of h into every u in the expression of f, we get the expression of the composite function $f(\frac{3x-2}{2})$, which is $f \bullet h$.

So putting h into u in $f(u)$, we get $f(h)$, which is the composite function: $(f \bullet h)(x)$.

That is, we get: $f(\frac{3x-2}{2}) = (f \bullet h)(x) = f\{h(x)\}$.

And we have already secured the expression of f, which is $3u + 4$, so we can just put the expression of h into u in $3u + 4$, then we get the expression of the composite function $f(\frac{3x-2}{2})$.

And thus, putting $\frac{3x-2}{2}$ into u in $3u + 4$, we get: $3 \cdot \frac{3(3x-2)}{2} + 4 = \frac{9x-6+8}{2} = \frac{9x+2}{2}$.

So we get: $f(\frac{3x-2}{2}) = \frac{9x+2}{2}$ for x real.

Note:

If not specified with a function, the domain is assumed to be a set of all real numbers or the largest set of real numbers the function can be defined.

Suggestions or Solutions

To the **Problem** in the **Example 1**

Assuming $f\left(\frac{2x-1}{x}\right) = (1-2x^2)(3x^2+x+2)$, find $f(8)$.

First, we can get: $8 = \frac{2x-1}{x} \Rightarrow 8x = 2x - 1 \Rightarrow 6x = -1 \Rightarrow x = -\frac{1}{6}$.

So next, we can get:

$$x = -\frac{1}{6} \Rightarrow (1-2x^2)(3x^2+x+2) = \left\{1 - 2\cdot\left(-\frac{1}{6}\right)^2\right\}\left\{3\cdot\left(-\frac{1}{6}\right)^2 + \left(-\frac{1}{6}\right) + 2\right\}$$

$$= \left(\frac{36}{36} - \frac{2}{36}\right)\left(\frac{3}{36} - \frac{6}{36} + \frac{72}{36}\right) = \frac{34}{36}\cdot\frac{69}{36} = \frac{17}{18}\cdot\frac{13}{12} = \frac{221}{216}. \quad \text{And thus, we get: } f(8) = \frac{221}{216}.$$

If not quite sure of the idea behind the processes above, follow the steps below:

Suppose first, $y = f(u)$, and $y = g(x) = \frac{2x-1}{x}$. Then, we get:

$y = (f\bullet g)(x) = f\{g(x)\} = f\left(\frac{2x-1}{x}\right) = (1-2x^2)(3x^2+x+2)$, which is a composite function.

So g is the function blends into, and blends into the function f.

And therefore next, we put x in terms of u solving for x the equation where $u = \frac{2x-1}{x}$, and then, we want to get the expression of f, don't we?

Not necessarily.　Why not?

To begin with, we know 8 in $f(8)$ is an output in the function g, since g is the function blends into the function f. And thus, of course, 8 is an input in the function f.

So when $u = 8$ in $f(u)$, the output is $f(8)$. And thus also, $f(8)$ itself is an output of the composite function $f \bullet g$.   How come?

In the expression $(1 - 2x^2)(3x^2 + x + 2)$, where is the x from?

The expression above is the expression of the composite function $f \bullet g$, so the x is the very x in $g(x)$, because g is the function blends into f, and we get the function $f \bullet g$.

So after such blending, we get the expression of $f \bullet g$, which is the expression above. So finding the x-value that makes the expression of g produce 8 as the output, what can we do with the x-value?

Suppose now, s is the x-value, that is, an input of g, and t is the output for the input s.

Then, we get a flow as below:

$$s \dashrightarrow x \text{ in } g(x) \dashrightarrow g(s) = 8 \dashrightarrow u \text{ in } f(u) \dashrightarrow f(8) = t.$$

And also, $f(8)$ is an output of the composite function $f \bullet g$.

For what input then, does the expression of $f \bullet g$ produce the output equal to the value of $f(8)$?
That is, for what x-value does $(1 - 2x^2)(3x^2 + x + 2)$ produce the value the same as $f(8)$?

It's s, of course, which is the x-value used in g when the expression of g produces the output 8.

So finding s, then putting s into x in the expression $(1 - 2x^2)(3x^2 + x + 2)$, we get t, which is the very value of $f(8)$. How can we find the x-value s, though?

The s is the x-value for which the output is 8 in the function g.

So the solution to $8 = \frac{2x-1}{x}$ is the x-value we want. And solving the equation, we get:

$8 = \frac{2x-1}{x} \Rightarrow 8x = 2x - 1 \Rightarrow 6x = -1 \Rightarrow x = -\frac{1}{6}$, which is the x-value s we want,

Therefore, we get: $g = 8$ when $x = -\frac{1}{6}$.

That is, $g(-\frac{1}{6}) = 8$. So next, 8 is an input of $f(u)$, which is the function g blends into.

Thus, as explained above, we get: $f(8) = (f \bullet g)(x)$ when $x = -\frac{1}{6}$. That is, $f(8) = (f \bullet g)(-\frac{1}{6})$.

So putting the x-value, $-\frac{1}{6}$ into x in the expression of $f \bullet g$, we get the value of $f(8)$.

And we have: $(f \bullet g)(x) = (1 - 2x^2)(3x^2 + x + 2)$. So now getting the value of $f(8)$, we get:

$x = -\frac{1}{6} \Rightarrow (1 - 2x^2)(3x^2 + x + 2) = \{1 - 2 \cdot (-\frac{1}{6})^2\}\{3 \cdot (-\frac{1}{6})^2 + (-\frac{1}{6}) + 2\} = (\frac{36}{36} - \frac{2}{36})(\frac{3}{36} - \frac{6}{36} + \frac{72}{36})$

$= \frac{34}{36} \cdot \frac{69}{36} = \frac{17}{18} \cdot \frac{13}{12} = \frac{221}{216}$.

And thus, we get: $f(8) = \frac{221}{216}$.

So finding the solution to an equation where $8 =$ the expression of g, that is, solving the equation: $8 = \frac{2x-1}{x}$, and then, putting the solution into x in the expression of $f \bullet g$, which is: $(1 - 2x^2)(3x^2 + x + 2)$, we get the value of $f(8)$.

Suggestions or Solutions
To the **Problem** in the Example 2

Assuming $f(x) = \frac{2x}{3x+1}$, and $g(x) = \frac{x}{3-2x}$, find the x-value that satisfies $(f \bullet g)(x) = f(x)$.

To begin with, $(f \bullet g)(x)$ is a composite function, where g blends into f.

Finding thus, $h = f \bullet g$ first, we get: $f \bullet g = f(g) = \frac{2g}{3g+1}$, where $g = \frac{x}{3-2x}$.

Meanwhile, $\frac{2g}{3g+1} = 2g \cdot \frac{1}{3g+1} = 2g \cdot (3g+1)^{-1}$.

So we get:

$2g = \frac{2x}{3-2x}$, $3g+1 = \frac{3x}{3-2x} + 1 = \frac{3x+3-2x}{3-2x} = \frac{x+3}{3-2x}$, and $\frac{2g}{3g+1} = \frac{2x}{3-2x} \cdot \frac{3-2x}{x+3} = \frac{2x}{x+3}$.

And thus, $(f \bullet g)(x) = \frac{2x}{x+3}$.

So next, $(f \bullet g)(x) = f(x) \Rightarrow \frac{2x}{x+3} = \frac{2x}{3x+1}$.

Thus, first, if $x = 0$, the equality holds. And next, assuming $x \neq 0$, and dividing both sides by $2x$, we get: $x + 3 = 3x + 1 \Rightarrow 2x = 2 \Rightarrow x = 1$.

$\therefore$ the x-value is **0** or **1**.

If not quite sure of the idea behind the processes above, follow the steps below:

This is nothing but an example of algebra practice on composite functions.

To begin with, assuming $h(x) = (f \bullet g)(x)$, we can say $h(x)$ is a composite function, where g blends into f.

Finding thus, $h = f \bullet g$ first, we get: $f \bullet g = f(g) = \frac{2g}{3g+1}$, where $g(x) = \frac{x}{3-2x}$.

Meanwhile, $\frac{2g}{3g+1} = 2g \cdot \frac{1}{3g+1} = 2g \cdot (3g+1)^{-1}$.

And thus, we have: $2g = \frac{2x}{3-2x}$, and $3g+1 = \frac{3x}{3-2x} + 1 = \frac{3x+3-2x}{3-2x} = \frac{x+3}{3-2x}$.

So we get: $\frac{2g}{3g+1} = \frac{2x}{3-2x} \cdot \frac{3-2x}{x+3} = \frac{2x}{x+3}$. And thus, we get: $h(x) = (f \bullet g)(x) = \frac{2x}{x+3}$.

And next, we want to find the x-value satisfying: $(f \bullet g)(x) = f(x)$.

That is to say that we want to solve the equation above.

First, we can set: $\frac{2x}{x+3} = \frac{2x}{3x+1}$. And thus, solving the equation, we get the x-value for which the equality can hold.

Solving it though, we may not want to just begin with dividing both sides by $2x$.

That's because x can be 0, and no division by 0 is allowed.

And in fact, if $x = 0$, the equality holds. So anyway, 0 is a solution.

Thus next, assuming $x \neq 0$, and dividing both sides by $2x$, we get:

$x + 3 = 3x + 1 \Rightarrow 2x = 2 \Rightarrow x = 1$.

And thus, 0 and 1 can be the x-value. That is, the x-value is 0 or 1.

And again, the domain matters. So we may want to note two cases below:

One is that in h, $x \neq -3$, since $h(x) = (f \bullet g)(x) = \frac{2x}{x+3}$, and the denominator cannot be 0.

And the other is that in h, x cannot be $\frac{3}{2}$, either.

That's because the domain of h is the domain of g since $h = f \bullet g$, so g is the function blends into, and the domain of g does not have $\frac{3}{2}$, since $g(x) = \frac{x}{3-2x}$, and the denominator cannot be 0.

So doing problems, we may want to check to see if the solution is in the domain.

And in particular, working with fractional expressions and radicals as $\sqrt[n]{x-1}$ where the index n is even as 2, 4, or 6, we want to make sure the denominator is not 0 or what's inside the radical sign is ≥ 0.

In short:

To begin with, $(f \bullet g)(x)$ is a composite function, where g blends into f.

Finding thus, $h = f \bullet g$ first, we get: $f \bullet g = f(g) = \frac{2g}{3g+1}$, where $g = \frac{x}{3-2x}$.

Meanwhile, $\frac{2g}{3g+1} = 2g \cdot \frac{1}{3g+1} = 2g \cdot (3g+1)^{-1}$.

So we get:

$2g = \frac{2x}{3-2x}$, $3g+1 = \frac{3x}{3-2x}+1 = \frac{3x+3-2x}{3-2x} = \frac{x+3}{3-2x}$, and $\frac{2g}{3g+1} = \frac{2x}{3-2x} \cdot \frac{3-2x}{x+3} = \frac{2x}{x+3}$.

And thus, $(f \bullet g)(x) = \frac{2x}{x+3}$.

So next, $(f \bullet g)(x) = f(x) \Rightarrow \frac{2x}{x+3} = \frac{2x}{3x+1}$.

Thus, first, if $x = 0$, the equality holds. And next, assuming $x \neq 0$, and dividing both sides by $2x$, we get: $x + 3 = 3x + 1 \Rightarrow 2x = 2 \Rightarrow x = 1$.

$\therefore$ the x-value is **0** or **1**. (The symbol $\therefore$ means therefore.)

Examples 4 in Composite Functions

0.   Suppose *a* is constant, $f: x \longrightarrow 9ax - 5$, and $g: x \longrightarrow 7 - 2ax$. Then:

0.0.   Find the value of *a* for which $g \bullet f = f \bullet g$.

0.1.   Find a function *h* so that $f \bullet h = g$, using the value of *a* found in the problem 0.

0.2.   Find a function *h* so that $h \bullet g \bullet f = f$, using the value of *a* found in the problem 0.

1.   Suppose $y = f_1(x)$, $f_2 = f_1 \bullet f_1$, $f_3 = f_2 \bullet f_1$, $f_4 = f_3 \bullet f_1$, $\cdots$ $f_k = f_{k-1} \bullet f_1$.

And in the graph below, *a, b, c, d,* and *e* are constant, and the curve of f_1 is in black.

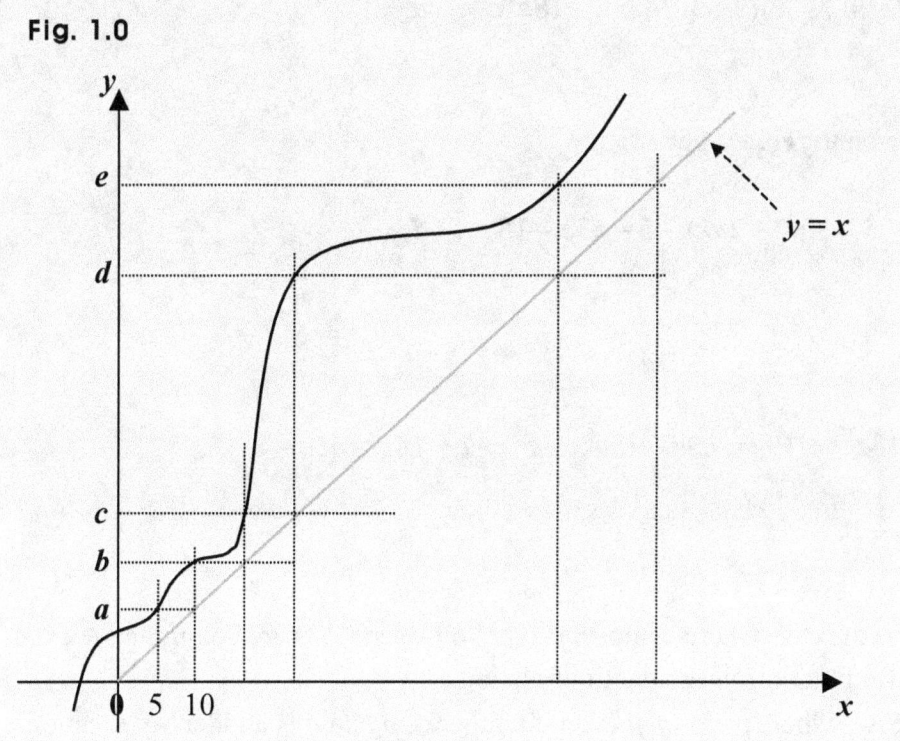

Fig. 1.0

Then:

0.   Find $f_4(10)$.

1.   Find *p, q,* and *r* so that we get: $f_2(p) = c, f_2(q) = b,$ and $f_3(r) = c$.

152

Suggestions or Solutions

Assuming a is constant, $f: x \longrightarrow 9ax - 5$, and $g: x \longrightarrow 7 - 2ax$, find the value of a for which $g \bullet f = f \bullet g$.

This is just another example of algebra practice on composite functions.

Now, we have: $y = f(x) = 9ax - 5$ for x real, and $y = g(x) = 7 - 2ax$ for x real.

So to begin with, finding $g \bullet f$, we get:

$g \bullet f = g(f) = 7 - 2af = 7 - 2a(9ax - 5) = 7 - 18a^2x + 10a$.

And next, moving on to $f \bullet g$, we get:

$f \bullet g = f(g) = 9ag - 5 = 9a(7 - 2ax) - 5 = 63a - 18a^2x - 5$.

Thus next, we get:

$g \bullet f = f \bullet g \Rightarrow 7 - 18a^2x + 10a = 63a - 18a^2x - 5 \Rightarrow 7 + 10a = 63a - 5$

$\Rightarrow 53a = 12 \Rightarrow a = \frac{12}{53}$.

And thus, it can be the case where composite-function-operations are commutative even if neither of the two functions is an identity function.
So we can have cases where $p \bullet q = q \bullet p$ even if neither of p and q is an identity function.

In other words, it is *sometimes* the case where $p \bullet q = q \bullet p$ if p and q are functions, of course. So it is not always the case.

  • Thus, *in general*, assuming F and G are functions, we have: $F \bullet G \neq G \bullet F$.

Suggestions or Solutions
To the **Problem 1** in the **Example 0**

Assuming a is constant, $f: x \longrightarrow 9ax - 5$, and $g: x \longrightarrow 7 - 2ax$, find a function h so that we get: $f \bullet h = g$, using the value of a found in the problem 0.0.

To begin with: $f(x) = 9 \cdot \frac{12}{53} x - 5 = \frac{108}{53} x - 5$, and $g(x) = 7 - 2 \cdot \frac{12}{53} x = 7 - \frac{24}{53} x$.

So first, we can get: $f(x) = \frac{108}{53} x - 5 \Rightarrow f(h) = \frac{108}{53} h - 5$. And next, we can get:

$\frac{108}{53} h - 5 = 7 - \frac{24}{53} x \Rightarrow \frac{108}{53} h = 12 - \frac{24}{53} x \Rightarrow h = \frac{53}{108}(12 - \frac{24}{53} x) = \frac{53}{12 \cdot 9}(12 - \frac{24}{53} x) = \frac{1}{9}(53 - 2x)$.

Therefore, $h(x) = \frac{1}{9}(53 - 2x)$ for x real.

If not quite sure of the idea behind the processes above, follow the steps below:

(Again, this is just another example of algebra practice on composite functions.)

Now, we have $f(x) = 9ax - 5$ for x real, and $g(x) = 7 - 2ax$ for x real.

And also, we found in the problem 1, that $a = \frac{12}{53}$. So now we have:

$f(x) = 9 \cdot \frac{12}{53} x - 5 = \frac{108}{53} x - 5$, and $g(x) = 7 - 2 \cdot \frac{12}{53} x = 7 - \frac{24}{53} x$.

Then, to begin with, we can put $f \bullet h = g$ this way, too: $f(h) = g$.

And next, we have: $f(x) = \frac{108}{53} x - 5$. So we get: $f(h) = \frac{108}{53} h - 5$.

And next, we have: $g(x) = 7 - \frac{24}{53} x$. And we have: $f(h) = g$, too.

So we can set: $\frac{108}{53} h - 5 = 7 - \frac{24}{53} x$. And thus, we get:

$\frac{108}{53} h - 5 = 7 - \frac{24}{53} x \Rightarrow \frac{108}{53} h = 12 - \frac{24}{53} x \Rightarrow h = \frac{53}{108}(12 - \frac{24}{53} x) = \frac{53}{12 \cdot 9}(12 - \frac{24}{53} x) = \frac{1}{9}(53 - 2x)$.

Therefore, we get: $h(x) = \frac{1}{9}(53 - 2x)$ for x real.

Suggestions or Solutions
To the **Problem 2** in the Example **0**

Assuming *a* is constant, *f*: *x* ⟶ 9*ax* – 5, and *g*: *x* ⟶ 7 – 2*ax*, find a function *h* so that we get: *h•g•f* = *f*, using the value of *a* found in the problem 0.0.

To begin with: $f(x) = \frac{108}{53}x - 5$, and $g(x) = 7 - \frac{24}{53}x$.

And setting: $m = \frac{108}{53}$, and $n = -\frac{24}{53}$, we have: $f(x) = mx - 5$, and $g(x) = nx + 7$.

So assuming $p = g•f$, we get: $p = g•f = g(f) = nf + 7 = n(mx - 5) + 7 = mnx - 5n + 7$.

So: $h(g•f) = h(p) = h(mnx - 5n + 7) \Rightarrow h(g•f) = f \Rightarrow h(mnx - 5n + 7) = mx - 5$.

And thus, assuming $y = h(t)$, we can set: $t = mnx - 5n + 7$.

So we get: $t = mnx - 5n + 7 \Rightarrow x = \frac{1}{mn}(t + 5n - 7)$. And we have: $f(x) = mx - 5$.

So we get: $m\frac{1}{mn}(t + 5n - 7) - 5 = \frac{1}{n}(t + 5n - 7) - 5 = \frac{1}{n}(t - 7) = \frac{24}{53}(7 - t)$, since $n = -\frac{24}{53}$.

Thus, we get: $h(t) = \frac{24}{53}(7 - t)$ for *t* real.

And putting it in the *x-y* system, we can just set: $y = h(x) = \frac{24}{53}(7 - x)$ for *x* real.

If not quite sure of the idea behind the processes above, follow the steps below:

When we get solutions, understanding matters, of course, but is just one thing. Doing it is another. Doing it, we do algebra. So algebra matters. It actually connects problems to solutions. And thus, this is another example of algebra practice on composite functions.

Now, doing this problem, we may want to use a tool as follows:

h•(g•f) = *(h•g)•f* = *h•g•f*, which is formally called a theorem.

So we may want to take advantage of it.

And we have: $f(x) = \frac{108}{53}x - 5$, and $g(x) = 7 - \frac{24}{53}x$.

Making it a bit simpler though, we can set: $m = \frac{108}{53}$, and $n = -\frac{24}{53}$.

Then, we have: $f(x) = mx - 5$, and $g(x) = nx + 7$.

So assuming $p = g \bullet f$, we get: $p = g \bullet f = g(f) = nf + 7 = n(mx - 5) + 7 = mnx - 5n + 7$.

Thus, we get: $p(x) = mnx - 5n + 7$.

So next, we get: $h(g \bullet f) = h(p) = h(mnx - 5n + 7)$.

And thus, we get: $h(g \bullet f) = f \Rightarrow h(mnx - 5n + 7) = mx - 5$.

Now, what are we after, though?

We are in fact, looking for the expression of h.

Then, we can notice that this is quite similar to the ones in **Examples 3**.

Assuming $y = h(t)$, we can set: $t = mnx - 5n + 7$.

That's because we have: $h \bullet p = h(p) = f$, and $p(x) = mnx - 5n + 7$.

And thus, solving $t = mnx - 5n + 7$ for x, we can put x in terms of t, that is, we get an expression in terms of t, and the expression equals to x.

Then, putting the expression into x in the expression of f, which is: $mx - 5$, we get the expression in the function h. If not quite sure, you may want to refer to **Examples 3**.

So now, solving $t = mnx - 5n + 7$ for x, we can simply get:

$t = mnx - 5n + 7 \Rightarrow x = \frac{1}{mn}(t + 5n - 7)$.

Thus next, putting $\frac{1}{mn}(t+5n-7)$ into x in the expression of f, we get:

$$m\frac{1}{mn}(t+5n-7)-5=\frac{1}{n}(t+5n-7)-5=\frac{1}{n}(t-7)=\frac{24}{53}(7-t) \text{ since } n=-\frac{24}{53}.$$

So we get: $h(t)=\frac{24}{53}(7-t)$ for t real.

And if we want to put it in the x-y system, we can just set: $y=h(x)=\frac{24}{53}(7-x)$.

We can use any letters as variables insofar as their relation is maintained.

In short:

We have: $f(x)=\frac{108}{53}x-5$, and $g(x)=7-\frac{24}{53}x$.

And setting: $m=\frac{108}{53}$, and $n=-\frac{24}{53}$, we have: $f(x)=mx-5$, and $g(x)=nx+7$.

So assuming $p=g\bullet f$, we get: $p=g\bullet f=g(f)=nf+7=n(mx-5)+7=mnx-5n+7$.

So we get: $h(g\bullet f)=h(p)=h(mnx-5n+7) \Rightarrow h(g\bullet f)=f \Rightarrow h(mnx-5n+7)=mx-5$.

And thus, assuming $y=h(t)$, we can set: $t=mnx-5n+7$.

So solving the equation for x, we get: $t=mnx-5n+7 \Rightarrow x=\frac{1}{mn}(t+5n-7)$.

Thus next, putting $\frac{1}{mn}(t+5n-7)$ into x in the expression of f, we get:

$$m\frac{1}{mn}(t+5n-7)-5=\frac{1}{n}(t+5n-7)-5=\frac{1}{n}(t-7)=\frac{24}{53}(7-t) \text{ since } n=-\frac{24}{53}.$$

So we get: $h(t)=\frac{24}{53}(7-t)$ for t real.

And putting it in the x-y system, we can just set: $y=h(x)=\frac{24}{53}(7-x)$ for x real.

Suggestions or Solutions
To the **Problem 0** in the Example **1**

Suppose $y = f_1(x)$, $f_2 = f_1 \bullet f_1$, $f_3 = f_2 \bullet f_1$, $f_4 = f_3 \bullet f_1$, $\ldots$ $f_k = f_{k-1} \bullet f_1$.

And in the graph below, a, b, c, d, and e are constant, and the curve of f_1 is in black.

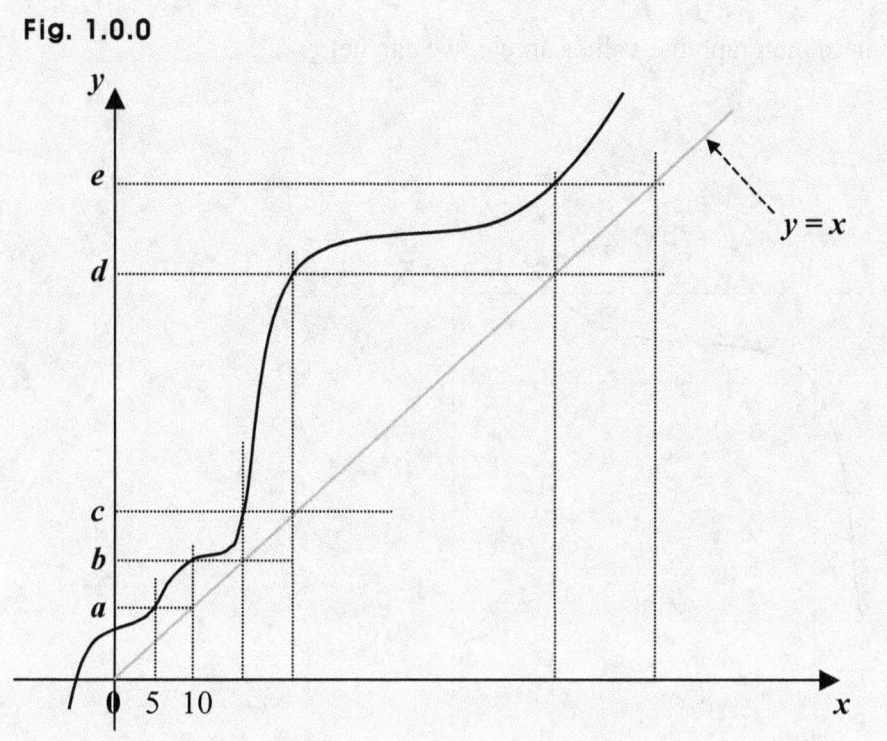

Fig. 1.0.0

Then, find $f_4(10)$.

$$f_4(10) = f_3(b) = f_2\{f_1(b)\} = f_2(c) = f_1\{f_1(c)\} = f_1(d) = e \Rightarrow f_4(10) = e.$$

If not quite sure of the idea behind the processes above, follow the steps below:

In this problem, we are given a sequence of composite functions.
So each term in the sequence is a composite function.
Each composite function is however, a blend of the same functions.
And the same function is f_1, used over and over again.
The first term is not though, a composite function but just a function $y = f_1(x)$.

Then, the second term is: $f_2 = f_1 \bullet f_1$, the third is: $f_3 = f_1 \bullet f_1 \bullet f_1$, and so forth.

Now, at each and every point the line gray above, both coordinates are the same. That is, x-coordinate = y-coordinate at every point in the line.

So we can notice that $a = 10, f_1(10) = b, f_1(b) = c, f_1(c) = d,$ and $f_1(d) = e$.

And thus, specifying in the graph the values above, we can get:

Fig. 1.0.1

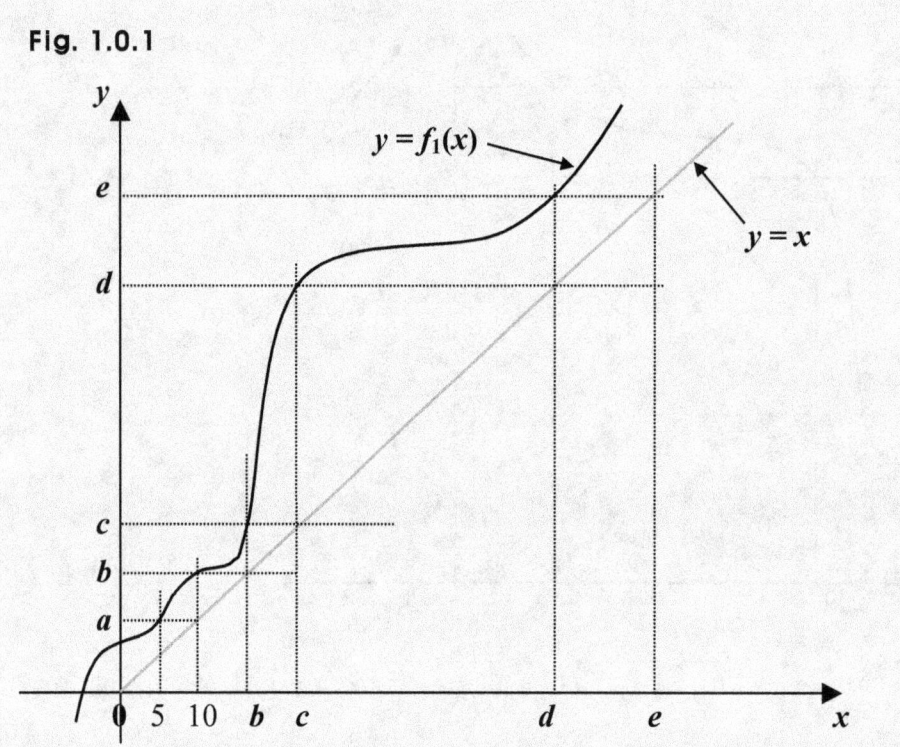

What we want to find is though, the output of the composite function $f_4(x)$ when $x = 10$. The curve and the values in the graph above are however, for the function f_1.

How then, can we get the output for the input 10 of the function $f_4(x)$?

We can use the sequence given.   Then, first, we can see that $f_4 = f_3 \bullet f_1$.

So we get: $f_4(x) = (f_3 \bullet f_1)(x) = f_3\{f_1(x)\} \Rightarrow f_4(10) = f_3\{f_1(10)\}$.

And in the graph above, we have: $f_1(10) = b$. So we get: $f_4(10) = f_3\{f_1(10)\} = f_3(b)$.

What then, about $f_3(b)$?

Going back to the sequence, we have: $f_3 = f_2 \bullet f_1$.

So we get: $f_3(x) = (f_2 \bullet f_1)(x) = f_2\{f_1(x)\} \Rightarrow f_3(b) = f_2\{f_1(b)\}$.   What then, about $f_1(b)$?

Going back to the graph, we can see that: $f_1(b) = c$.

So we get: $f_3(b) == f_2\{f_1(b)\} = f_2(c)$.

And next, going back to the sequence, we get: $f_2 = f_1 \bullet f_1 \Rightarrow f_2(c) = f_1\{f_1(c)\}$.

Then again, backing to the graph, we can get: $f_1(c) = d$.

So we get: $f_2(c) = f_1\{f_1(c)\} = f_1(d)$.

Then again, backing to the graph, we can get: $f_1(d) = e$.

So putting threads together, we get:

$f_4(10) = f_3(b) = f_2\{f_1(b)\} = f_2(c) = f_1\{f_1(c)\} = f_1(d) = e$. And thus, $f_4(10) = e$.

In short:

$f_4(10) = f_3(b) = f_2\{f_1(b)\} = f_2(c) = f_1\{f_1(c)\} = f_1(d) = e$.

Suggestions or Solutions
To the **Problem 1** in the Example **1**

Suppose $y = f_1(x)$, $f_2 = f_1 \bullet f_1$, $f_3 = f_2 \bullet f_1$, $f_4 = f_3 \bullet f_1$, $\ldots$ $f_k = f_{k-1} \bullet f_1$.

And in the graph below, a, b, c, d, and e are constant, and the curve of f_1 is in black.

Fig. 1.1.0

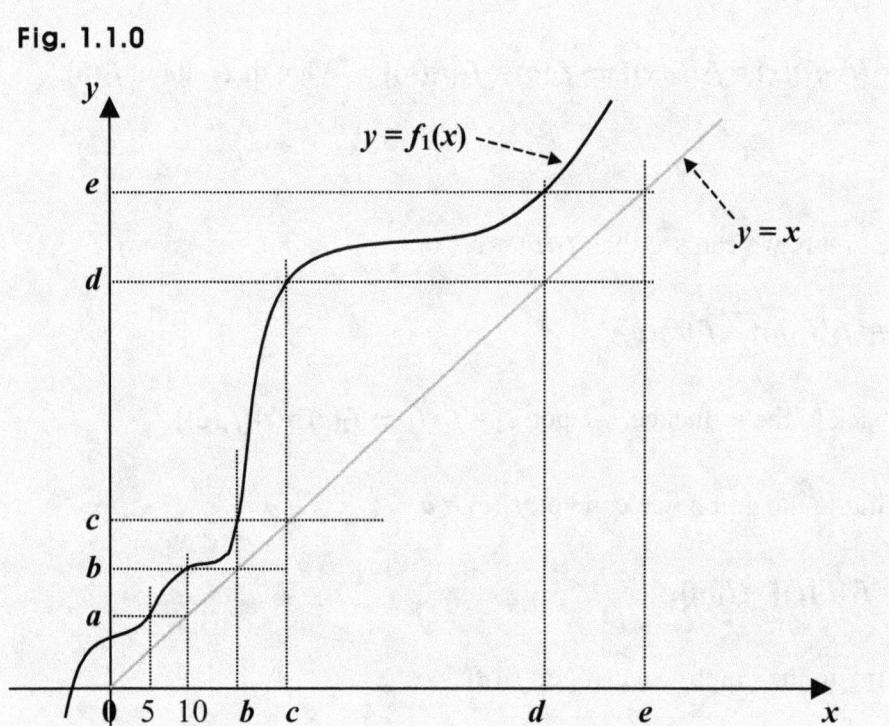

Then, find p, q, and r so that we get: $f_2(p) = c$, $f_2(q) = b$, and $f_3(r) = c$.

$f_2 = f_1 \bullet f_1 = f_1(f_1) \Rightarrow f_2(p) = f_1\{f_1(p)\} = c \Rightarrow f_1(p) = b$. And $f_1(10) = b$. So $p = 10$.

$f_2 = f_1 \bullet f_1 = f_1(f_1) \Rightarrow f_2(q) = f_1\{f_1(q)\} = b \Rightarrow f_1(q) = 10$, because $f_1(10) = b$.

And also, we get: $a = 10$, and we have: $f_1(5) = a = 10$. So $q = 5$.

$f_3 = f_2 \bullet f_1 = f_1 \bullet f_1 \bullet f_1 \Rightarrow f_3(r) = f_1[f_1\{f_1(r)\}] = c \Rightarrow f_1\{f_1(r)\} = b \Rightarrow f_1(r) = 10 \Rightarrow r = 5$.

If not quite sure of the idea behind the processes above, follow the steps below:

Beginning with the equation $f_2(p) = c$, we can use: $f_2 = f_1 \bullet f_1$, so we can put it this way:

$f_2(p) = (f_1 \bullet f_1)(p) = c$. What does this equation mean, though?

The equation is saying that from the composite function $f_2(x) = (f_1 \bullet f_1)(x)$, the output c gets produced for a particular value of x in f_2, and the equation is asking us to find *the particular value.*

We don't even know though, the expression of the function $f_2(x)$.

We know however, $f_2 = f_1 \bullet f_1 = f_1(f_1)$, so we can put it this way, too: $f_2(p) = f_1\{f_1(p)\} = c$.

And setting $u = f_1(p)$, we can put it this way, also: $f_2(p) = f_1(u) = c$.

Then, looking at the graph above, we can see that $u = b$.

So we get: $b = f_1(p)$. Thus again getting back to the graph above, we can see: $b = f_1(10)$.

That is, we get: $b = f_1(p) = f_1(10)$.   So we get: $p = 10$.

In short:

$f_2 = f_1 \bullet f_1 = f_1(f_1) \Rightarrow f_2(p) = f_1\{f_1(p)\} = c \Rightarrow f_1(p) = b$. And $f_1(10) = b$. So $p = 10$.

Let's next, move on to the equation where $f_2(q) = b$.

Then again, we have: $f_2 = f_1 \bullet f_1$, so we can put it this way: $f_2(q) = (f_1 \bullet f_1)(q) = b$.

And the equation is saying that from the composite function $f_2(x) = (f_1 \bullet f_1)(x)$, the output b gets produced for a particular value of x in f_2, and the equation is asking us to find *the particular value.*

162

Then again, since we have: $f_2 = f_1 \bullet f_1 = f_1(f_1)$, we can set: $f_2(q) = f_1\{f_1(q)\} = b$.

And setting $v = f_1(q)$, we can set: $f_2(q) = f_1\{f_1(q)\} = f_1(v) = b$.

Then, looking at the graph above, we can see that $f_1(10) = b$, so we get: $v = 10$.

So we get: $v = f_1(q) \Rightarrow 10 = f_1(q)$.   Then again in the graph, we can see that $a = 10$.

That's because at every point in the line $y = x$, both coordinates are the same.
Thus, we get: $f_1(q) = a$.

Then again, backing to the graph, we can see: $f_1(5) = a$.
Then, we get: $f_1(q) = f_1(5)$.   So we get: $x = 5$.

In short:

$f_2 = f_1 \bullet f_1 = f_1(f_1) \Rightarrow f_2(q) = f_1\{f_1(q)\} = b \Rightarrow f_1(q) = 10$, because $f_1(10) = b$.

And also, we get: $a = 10$, and we have: $f_1(5) = a = 10$. So $q = 5$.

And moving on to the next, we have to solve this: $f_3(r) = c$.

First, we have: $f_3 = f_2 \bullet f_1 = f_1 \bullet f_1 \bullet f_1$, so we can put it this way: $f_3(r) = f_1[f_1\{f_1(r)\}] = c$.

And examining the graph, we can see that $f_1(b) = c$.   So we get: $f_1\{f_1(r)\} = b$.
That's because: $f_3(r) = f_1[f_1\{f_1(r)\}] = f_1(b) = c$.

Then again, in the graph, we can see $f_1(10) = b$.   So we get: $f_1(r) = 10$.
That's because: $f_1\{f_1(r)\} = f_1(10) = b$.

And backing to the graph again, we can see: $f_1(5) = a$.

So we get: $f_1(5) = 10$, because $a = 10$. Thus, we get: $f_1(5) = f_1(r)$. So we get: $r = 5$.

In short:

$f_3 = f_2 \bullet f_1 = f_1 \bullet f_1 \bullet f_1 \Rightarrow f_3(r) = f_1[f_1\{f_1(r)\}] = c \Rightarrow f_1\{f_1(r)\} = b \Rightarrow f_1(r) = 10 \Rightarrow r = 5$.

www.ingramcontent.com/pod-product-compliance
Lightning Source LLC
Chambersburg PA
CBHW081121170526
45165CB00008B/2510